THE ISLAND OF VOLCANOES

A GUIDE TO LANZAROTE GEOLOGY AND LANDSCAPE

ROGER TREND

ediciones remotas

First published by:
Ediciones Remotas

Printer:
Editorial MIC

ISBN: 978-84-127773-2-1

Legal deposit: GC 505-2023

Second edition 2025

This book has been printed on FSC certified paper, with vegetable and mineral-based inks, free from heavy metals, and bound with elements that do not impede the recycling process.

CONTENTS

FOREWORD

The island of Lanzarote is of unquestionable natural beauty, which explains why it is so attractive to tourists and naturalists alike, who visit in the millions every year. Moreover, Lanzarote is a mysterious island full of secrets, especially in the field of geology. To begin with, it is not even a separate island. Although geographically and administratively it is, the Bocaina Strait, that separates it from Fuerteventura, was not always submerged. Over long periods of time, it was transformed from a simple valley to a shallow Strait due to rising sea levels after the last glacial maximum. In fact, during the last ice age (i.e. the glacial maximum), it would have been possible to cross on foot from neighbouring Fuerteventura, but we have no evidence that humans were living on the islands at that time. Geologically, however, the islands of Lanzarote and Fuerteventura are formed by an alignment of old basaltic shield volcanoes, that marked a chain of independent edifices that grew along a fracture that was parallel to the African continental margin. The most recent volcanism of Lanzarote, in turn, is what in geology is called post-shield or rejuvenation volcanism, which together with erosion and sedimentation assembled these old shields into an elongate landmass that stretches from Lanzarote to Fuerteventura with only a shallow drowned valley that separates the two islands today. Lanzarote's most spectacular feature is undoubtedly the most recent volcanism, however, produced in a great fissure-fed eruption that lasted from 1730 to 1736 and another smaller one in 1824. The lavas and ashes from these eruptions covered large portions of the island with a thick blanket of small black basaltic stones (lapilli), punctuated by a considerable number of volcanic cones, which gave the island the local name of the "island of a thousand volcanoes". This gives the impression of much of the island having formed during recent volcanic events, but in fact, Lanzarote is more like an old and dilapidated building that got a new coat of volcanic paint during the 1730-36 eruption, making it look as good as new. It is pure 'geological cosmetics', however, since the bulk of the island's volume is from the old Miocene shield volcanoes that are more than five million years old, with the most recent volcanism being just a thin cover on top of this much older substrate.

Indeed, chronicles from before the 18th century and prior to the Timanfaya eruption tell us that Lanzarote was a dry island, and farming focussed on cereal crops as the basis of most people's livelihood, sustaining a thinly spread population only. When the eruption started, worries in Madrid were that the eruption of 1730 would be the final nail in the coffin for the island's economy, but Lanzarote always surprises. Although initially a catastrophe, the eruption soon became a gift of nature to the islanders. The inhabitants, who had initially fled the island, noted on their return that, against all odds, on the fields where the lapilli layers were only a few centimetres thick, the vegetation took hold and developed splendidly, as if someone had magically watered the plants. The wide lapilli covered plains left by the eruption soon transformed into orchards where vegetables of all kinds could be harvested (e.g. potatoes, grapes, and dates) and the population of Lanzarote has been growing ever since, doubling already in the 50 years after the 1730-36 eruption. This was because the vesicles (bubbles) of the small volcanic stones (lapilli) acted like a sponge for the condensed moisture of the early hours of the day, preventing the frequent winds from drying out the upper layers of soil and ash. It is surprising to dig under the lapilli on a dry, sunny day and find that the soil remains moist, a feature that allowed agricultural yields to multiply in the decades after the Timanfaya events.

Lanzarote is also home to one of the most emblematic lava tubes on the planet. With a length of more than 7.5 kilometres and a diameter of up to 35 metres, the La Corona lava tube reaches all the way to the coast and continues offshore, i.e. below the seabed for another 1.6 kilometres to a depth of almost 70 metres. Researchers wondered for quite some time how this might be possible and the Corona lava tube presented a magnificent puzzle for Earth scientist. Indeed, it seems to be another trick of Lanzarote, because it is absolutely not possible that an active volcanic tube could reach such a length and depth under the ocean water. As so often with magic, it is pure appearance and again, during the last major glaciation, vast amounts of water were locked at the polar ice caps and the sea level of the Atlantic Ocean was significantly lower. In fact, it was so low that the Corona lava flowed until it reached the coast, which was where the volcanic tube now ends. Radiometric dating of the lava flow that hosts the tunnel yielded an age of 21,000 years before present, i.e. during a glacial maximum, and so the Corona lava mystery can now tell us where sea level was located at that time in the past. This twist of 'scientific' magic now helps us to better understand the world around us and the changes that occur in nature through time.

Lanzarote still has many more mysteries in store. Large ostrich eggs were found on a fossil Miocene beach preserved in between Miocene lava flows. But how was it possible for these birds to get to Lanzarote, since ostriches cannot fly? Early geologists solved this enigma by proposing a now vanished "continental land bridge" between Lanzarote and Africa, which would have allowed migration of ostriches on foot. But geology closed this door as it was discovered later that the islands are independent geological entities and that there was no land bridge between Lanzarote and the mainland, or between Lanzarote and the islands in the Central part or the West of the archipelago. The solution is as simple as it is surprising. The large bird eggs, which are as large as ostrich eggs, actually do not belong to a flightless bird, but to a now extinct large flying bird of the albatross family, who seemingly used the beaches of Lanzarote for nesting.

Many more surprising and fascinating aspects of Lanzarote's geology await the visitor and this is precisely what Roger Trend in his book "The Island of Volcanoes: a guide to lanzarote geology and landscape" allows us to explore. In 192 pages, splendidly illustrated with graphs and many spectacular photographs, the reader is taken onto a journey into the details of Lanzarote's geological origin and evolution, including the formation of the Ajaches and the Famara shield volcanoes that form the foundations of the island, but are deeply eroded now, allowing only the skilled visitor to read the history of growth, evolution and decay of these long lost large volcanic giants in the rock record. Their demise gave rise to a period of volcanic quiescence and widespread erosion of the island. The book then walks the reader through many of the younger 'rejuvenated' volcanic cones and vents, usually referred to as 'volcánes' in local terminology, although they are often more like larger vent complexes. The reader will explore Montaña Roja, Atalaya de Femés, Caldera Blanca, Risco Quebrada and Montaña Caldereta, Volcan de Corona, and El Gofo that all paint a vivid picture of Lanzarote's volcanic history after the large erosion episode that eventually gave rise to the disastrous Timanfaya eruption of 1730 to 1736 and the smaller historical eruption of 1824. In addition to exploring the various flows and flow fields, Trend's book also accompanies the reader into the spectacular lava tubes of Lanzarote before exploring the sedimentary deposits including the beautiful sand dunes and the contribution from blown-in Sahara dust to these sediments and dunes. However, the other side of volcanoes is also shown, the side that is often forgotten in light of volcanic hazards and violent eruptions, and that is the many resources volcanoes provide. From allowing us to use volcanic craters to produce sea salt to the widespread use of lapilli (picon) to grow wine and fruit, to inspiring modern art and contemporary architecture that integrates natural volcanic features with sophisticated modern architecture to melt geology and volcanoes with modern lifestyles and cultural sophistication. The volcanic hazards are thus balanced against the beauty and rich resources volcanoes provide, and which together create the sense of a deep link between natural forces from within the Earth with our ability to sculpt these natural landscapes to help us improve our lives on the surface of the planet. Everything is just that little bit clearer and easier to grasp on Lanzarote, and the mysteries of nature are therefore just that little bit closer to being solved on this stunning island. We all owe our gratitude to Roger Trend for taking us on this natural, scientific, and spiritual journey.

Prof. Juan Carlos Carracedo and Prof. Valentin R. Troll
(Earth Science Fellows of the Royal Academy of Sciences of the Canary Islands)

GRACIOSA
ALEGRANZA
Montaña Clara
GRACIOSA
LANZAROTE
LA BOCAYNA
LOBOS
OCÉANO ATLANTICO SEPTENTRIONAL.
CARTA
DE LA ISLA DE LANZAROTE
EN LAS CANARIAS,
Dirección de Hidrografía.
Madrid, 1852.
Corregida en 1868.
NOTAS

PREAMBLE

Lanzarote is stunning. It's one of the best places in the world to engage first hand with volcanoes. Why? The landscapes are unique and beautiful [**1**], the volcanoes and lava fields can be explored easily, the rocks are highly visible, the geological story can be summarised clearly, the island is easy to visit, the weather is great (usually!) and there is plenty of accommodation. And the whole package has been enhanced enormously by César Manrique.

Covering the main island [**2**], this guide recounts Lanzarote's geological history in a simplified way. It draws on recent research and shows how Lanzarote's rocks and landscapes give evidence of past geological activity. The main story starts with lava emerging on to the ocean floor some **70 million years ago (Ma)** and ends with the historical eruptions of 1730 - 1736 and 1824.

Much of Lanzarote's geological history can be understood from the rocks and landscapes, especially if we know what we are looking at. The evidence is out there: we can read the rocks. But some readers might not know the language, so some basic geology is covered as it crops up (**info boxes**).

(Previous spread) 1. Beautiful volcanic landscape from the top of Montaña Guardilama

◀ 2. Map of Lanzarote 1835 made by the English navy

1. ORIGINS

1.1. Planet Earth: continents and oceans

Lanzarote resulted from the inner workings of planet Earth [**3**]. Mainly. It's a layered planet [**4**] and the **rigid mantle** matters a lot, making up 84% of its volume. Towards its top the mantle includes a weak, partly-molten layer called the **asthenosphere**, around 100 km thick. Everything above the asthenosphere is the rigid **lithosphere**, typically 50 - 100 km thick. This comprises interlocking **tectonic plates** which slowly shuffle around on the weak asthenosphere. The lithosphere has two zones: **upper mantle** and (on top) **crust**. There are two types of lithosphere: **oceanic** and **continental**. Some plates move as fast as 9 cm a year but the Canaries, sitting on the Atlantic bit of the African Plate, move about 2 cm a year towards the north east. Most events outlined below are explained by the mega-theory of **plate tectonics**.

Around 300 Ma the world's continents were clustered as one supercontinent – **Pangaea**, surrounded by one super-ocean – **Panthalassa**. During the Triassic Period [5] Pangaea started to fragment, with Africa departing from North America. New ocean plate lithosphere was (and still is) created along the rift, the upwelling **magma** cooling to create **igneous** ("fire-formed") rock. This initiated the Mid-Atlantic Ridge, a submarine and highly active volcanic mountain range rising some couple of thousand metres above the ocean floor. These new **oceanic plates**, initially 5 to 10 km thick, are made of dark, dense igneous rocks, mainly **basalt**, **gabbro** and **peridotite**. They get thicker with age and also accumulate sedimentary rocks on their backs over millions of years as they trundle along: see ***Igneous rocks and lava*** info box.

As the African plate (part ocean, part continent) shifts north-eastwards, Atlantic's ocean lithosphere gradually becomes older, colder and denser, so it slowly subsides. And it has become progressively covered by an increasing thickness of edimentary rocks (eg limestone, sandstone) over those 180 **million years** (**Myr**). The Canaries sit on this dense, old, cold plate. Tectonic plates are not dragged along by a convecting mantle beneath, as was once thought. They shift by a combination of **push** from the higher and hotter mid-ocean ridges and **pull** by the lower and colder subduction zones. It's gravity at work.

◀ 3. Lanzarote's lava shows the inner workings of the Earth.
We see in this aerial photograph petrified lava measuring 20 metres from left to right.

EARTH STRUCTURE BENEATH THE CANARY ISLANDS
(not to scale)

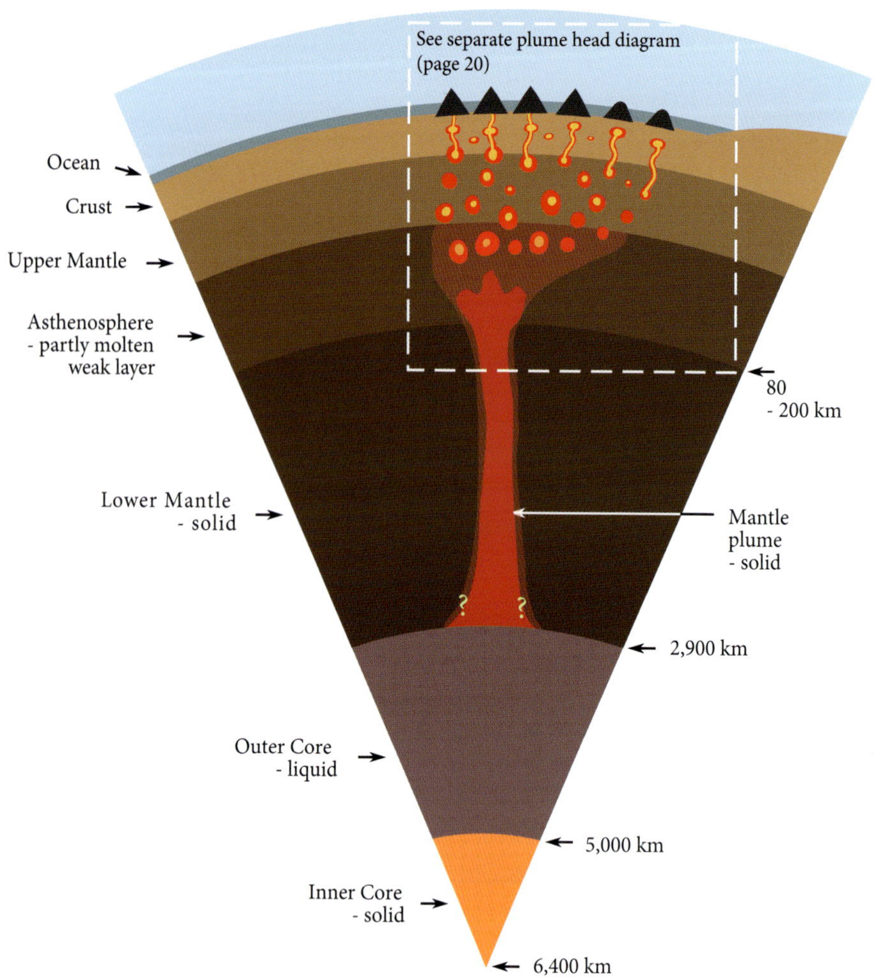

4. Earth - internal structure

1.2. Magma and mantle plumes

To melt rock you have to raise its temperature or reduce its pressure. **Magma** is normally generated within the asthenosphere where temperatures and pressures are occasionally just right for melting – Goldilocks conditions. Most of the mantle is solid rock because the pressure is too great to allow melting, despite the high temperatures. Most of the lithosphere is solid rock too, but for different reasons: those rocks are too cool to melt, despite the low pressures. Only in the asthenosphere are **pressures just low enough and temperatures just high enough** for rocks to exist close to their melting points. All they need is a bit lower pressure or a bit higher temperature (or both): and the magma's away, slowly rising into the lithosphere because of its buoyancy. Small magma blobs feed larger blobs, some of which rise into the crust, occasionally making it to the surface as volcanism. The presence of water in the magma also affects its melting: more water means easier melting.

Canary Island magma starts its life as solid rock near the base of the Earth's mantle at around 2,000 to 3,000 km. Over millions of years a **mantle plume** of **hot rock** rises through the mantle, heated mainly by radioactive decay. It eventually widens and partly melts into **magma blobs** which ascend within the asthenosphere and lower lithosphere [**9**]. These accumulate within the lithosphere as large magma bodies, much of which cools, solidifies and stays at those depths as giant domes of solid rock. However, some fluid magma (of basalt composition) pierces the lithosphere to feed volcanoes, creating the volcanic region of the Canaries (this is present tense because it's still happening). The rock's final melting is triggered by exceptionally

GEOLOGICAL TIMESCALE: THE LAST 250 MILLION YEARS

Period	Epoch	Date
Quaternary	Holocene	11,700 yrs
	Pleistocene	2.6 Ma
Neogene	Pliocene	5.3 Ma
	Miocene	23 Ma
Palaeogene	Oligocene	34 Ma
	Eocene	56 Ma
	Palaeocene	66 Ma
Cretaceous	Upper/Lower	145 Ma
Jurassic	Upper/Middle/Lower	200 Ma
Triassic	Upper/Middle/Lower	250 Ma
A further 4,400 million years of Earth history		

5. Geological timescale

IGNEOUS ROCKS AND LAVA

Magma is hot molten rock which originates in the Earth's interior. When it emerges on to Earth's surface it's called **lava**, regardless of its chemical make-up and whether or not it has cooled and solidified to form rock. When magma cools slowly within the crustal rocks it produces **igneous** rocks with large crystals, many larger than 10 mm. When lava cools rapidly on the surface it produces igneous rocks with small crystals, typically smaller than 1 mm. **Basalt** is an example. Most of Lanzarote's igneous rocks are basalt. **Vesicular** lava contains gas bubbles (vesicles) formed during eruption. Lava flows with crusts forming their surfaces are called pahoehoe, often "ropy" in character [**6**]. Lava flow with fragmented surfaces which move with the flow are called aa [7], usually blocky in character. Both labels are Hawaiian.

6. Pahoehoe lava

7. Aa lava

high temperatures plus the rapidly reducing pressure. Canaries magma is drier than normal because of its deep origins, although (sea)water has a crucial role in some eruptions but only at shallow depths. These mantle plumes (**hot spots**) are long-lived and stationary, which explains their large-scale and long-term supply of magma. More locally, it is the movement of magma at shallower depths beneath Lanzarote, its volcanic plumbing, which explains in detail the island's various volcanic episodes.

1.3. Canaries

Probably because of a large mantle plume, the Canaries started to develop as submarine piles of lava on the Atlantic floor around 70 Ma, creating seamounts (islands yet to emerge above the waves) on the ever-widening Atlantic floor. The first of these to form dry land created the East Canary Ridge: Lanzarote and Fuerteventura [**8**]. The oldest exposed rocks (21 Ma) are on Fuerteventura, Lanzarote first emerging 15 Ma. The other Canaries broke the waves progressively over the following 20 Myr as the African plate moved over the stationary plume towards the north east (it's still moving). Consequently, the youngest islands are to the west: El Hierro (1.1 Myr) and La Palma (1.7 Myr).

CANARY ISLANDS

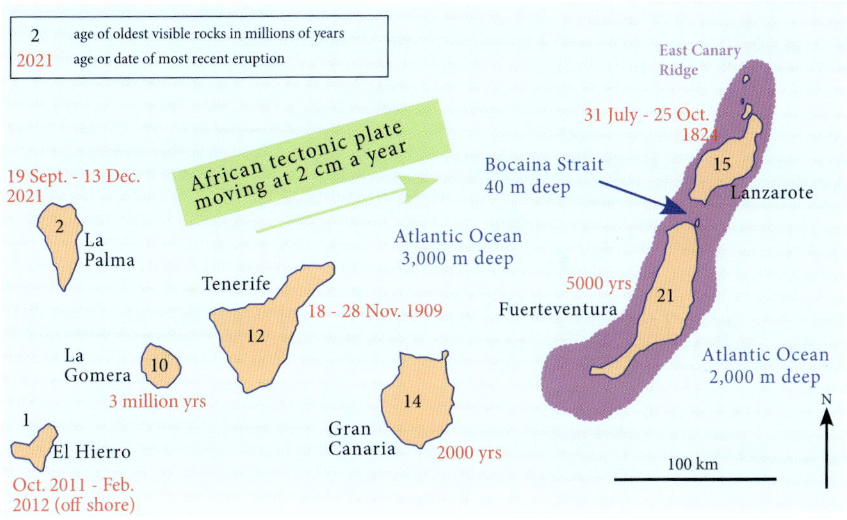

8. Canary Islands

CANARY ISLANDS: MANTLE PLUME HEAD FEEDING MAGMA BODIES
(not to scale)

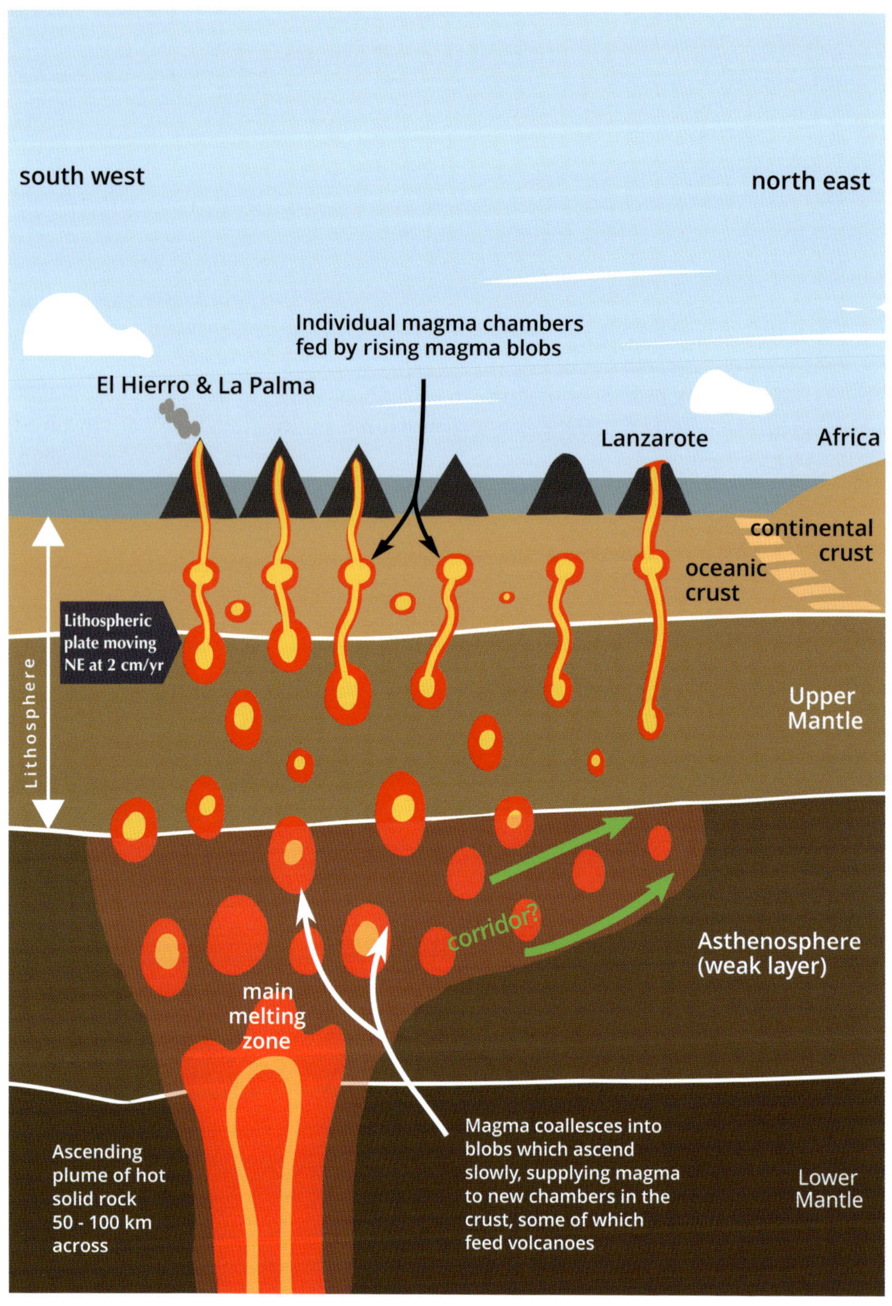

9. Canaries mantle plume head

The most recent Canaries eruption started on 19 September 2021 at the Cumbre Vieja ridge [**11**] on La Palma and ended on 13 December, with official cessation announced on Christmas Day. By that time around 7,000 people had been evacuated, more than 3,000 buildings damaged or destroyed and 1250 hectares (over 3,000 acres) buried, much of it valuable farmland. It was the most destructive eruption in Spain's history because of its location in a well-populated and cultivated region. In July 2022 the new volcano was (democratically) named Tajogaite, the Guanche name for Montaña Rajada. La Palma's recent volcanic eruptions and products are highly revealing, being similar (but not identical) to those in Lanzarote's past (see GOING FURTHER).

Plumes are not everything. Fifty years ago most theories on Canary Island origins emphasised the role of deep fracture zones within the lithosphere, zones of weakness caused by huge forces within tectonic plates as they shuffle along. But here's our first problem: volcanism normally occurs at random points along these fracture zones, not progressively from one end to the other as with the Canaries. So, the idea of a stationary plume piercing a moving plate has been the favoured explanation for a few decades, but it is likely that both plume and fracture zones are involved, the plume supplying magma and the NE/SW fractures determining volcano alignments. Research continues.

We have a second problem. Lanzarote has moved about 450 km away from the stationary hot spot, now centred beneath El Hierro and La Palma, so how did Lanzarote's recent (1730's) magma arrive? Briefly, this might be normal plume behaviour, or possibly some of the plume material may have been diverted towards the north east in a horizontal corridor within the asthenosphere, possibly even travelling beyond the Atlas Mountains of North Africa [**9**]. Magma blobs arose from this corridor to feed Lanzarote's more recent eruptions. Research continues here too.

2. FOUNDATIONS

2.1. The south: Los Ajaches Volcano

During the Miocene Epoch, around 15 Ma, **Hawaiian-type eruptions** produced a large **shield volcano** forming an island centred around Las Breñas and Femés in the south. Fluid basalt lava emerged at around 1,000 ^{0}C from a central vent and vertical cracks (**dykes**) within the growing volcano: see ***Dykes and Hawaiian eruptions*** info box. This was **Los Ajaches Volcano**, passing through 3 stages and eventually reaching 1.5 km in height [**27**].

2.1.1. Stage 1: Shield-building

After building up the submarine volcanic pile from 70 Ma, the **shield building stage** involved rapid and voluminous basaltic outpourings from 15 to 14 Ma and again from 13 to 12 Ma [**13**]. Less lava was ejected after 12 Ma, with intermittent activity occurring over the following millions of years. Lava emerged from a central vent, smaller secondary vents and many dykes. Beautiful examples of Miocene dykes can be seen around Papagayo in the south, especially along the coast where they can be examined easily [**12**]. Hawaiian eruptions are occasionally explosive, producing tephra, the name for all the particles which become airborne during an eruption: see ***Pyroclastic rocks and tephra*** info box.

Los Ajaches Hills [**10**] contain two **Geosites**. These are sites of special geological interest identified as part of Lanzarote's Global Geopark designation by UNESCO (see later). One relates to the Papagayo dykes and the second to the structures formed by gaps in Miocene lava eruptions: see ***Geosites*** info box.

2.1.2. Stage 2: Erosion

Ajaches' main **erosional stage** ran from 6 to 2.3 Ma. The hills were dissected into steep valleys and ridges running south-eastward to the sea where the pile of gently dipping Miocene lavas can now be viewed south from Playa Quemada [**14**]. Erosion

(Previous spread) 10. View of the village Femés and Los Ajaches

◀ 11. La Palma eruption 2021

DYKES AND HAWAIIAN ERUPTIONS

Named after Hawaii, these eruptions are relatively quiet with repeated outpourings of fluid (runny) basalt lava producing gentle slopes. Usually they erupt intermittently over many years, centuries, millennia or longer, so they produce polygenetic volcanoes (formed by repeated eruption events). The result is a **shield volcano,** shaped like a warrior's shield placed on the ground, convex surface up [**13**]. Several main vents may develop, but usually most lava emerges from vertical **dykes.** These are sheets of magma squeezed along (roughly vertical) cracks, cutting across the existing (roughly horizontal) lava flows and often radiating from the volcano's centre. Magma rises within a dyke, emerging as a fissure eruption to add lava to the expanding shield volcano [**12**]. As the eruption dies down, magma solidifies within the dyke to form a solid **dolerite** sheet (dolerite is an igneous rock chemically like basalt but with larger crystals because of its slower cooling). Lanzarote dykes range in thickness from a few centimetres to several metres, landscape erosion often leaving them as low ridges because of their toughness (dyke is an old Scottish word for a low stone wall).

12. Papagayo dykes

LANZAROTE'S THREE TYPES OF UPLAND

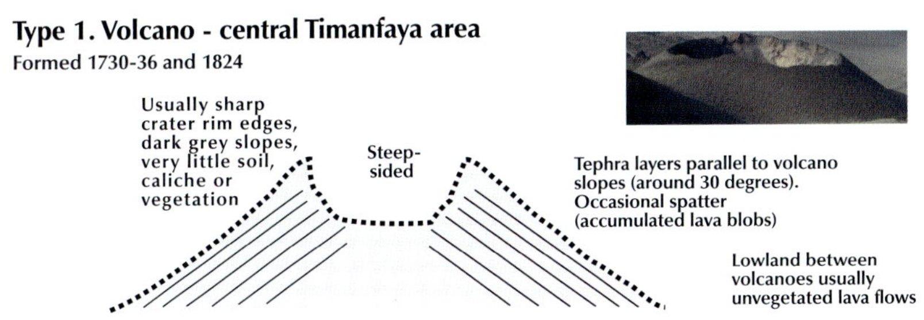

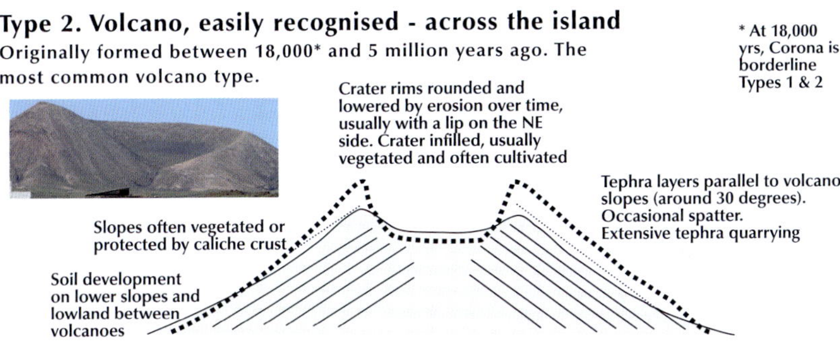

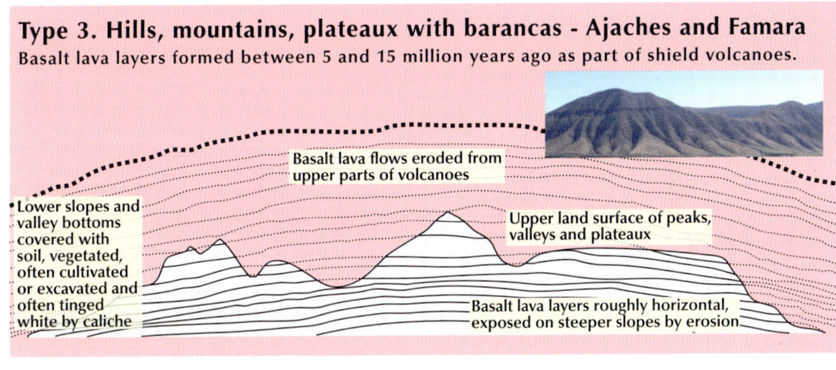

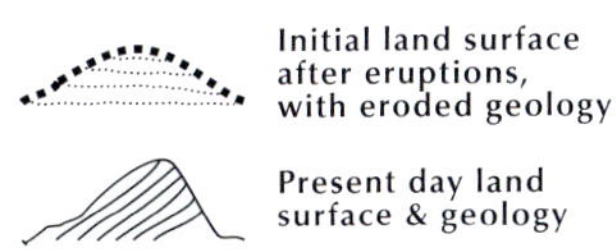

13. Three types of upland, not to scale

14. Ajaches near Playa Quemada

15. Yaiza hillside

PYROCLASTIC ROCKS AND TEPHRA

When volcanic material is thrown into the air during an explosive eruption, lava droplets and rock particles collect as particles around the vents. This (usually airborne) **tephra** behaves as sediment if it solidified while airborne: blown by wind, carried by water, sorted by particle size and deposited in layers. Rocks formed by this combination of volcanic and sedimentary processes are called **pyroclastic** (**pyro** means fire and **clastic** means fragments). Tephra is classified by particle size: **ash** grains are smaller than 2 mm, **lapilli** between 2 and 64 mm and **blocks** are larger than 64 mm. Fully airborne blocks are called bombs. Blocks full of **vesicles** (gas bubbles) are called scoria. In the Canaries lapilli are called 'picon' (also 'rofe') and ash is called 'arena'. When ash gets lithified into solid rock it's called **tuff**.

on the west removed much of the volcano, leaving a steep scarp [**20**] where these layered basalts and pyroclastic rocks are revealed by erosion in the hillsides around Yaiza and Femés [**15**]. Erosion has produced two broad valleys running north-eastwards towards Uga and Yaiza: Femés and Fena Valleys respectively [**20**]. Both are rift valleys (graben) formed as the land shifted downwards between parallel faults.

During this erosional stage there was time for soils and white **caliche** to develop: see ***Caliche*** info box. Sediments in shades of brown and orange are produced by the weathering of basalt, reflecting the oxidation of the rock's iron minerals. Basically it's rust, iron oxide [**16**]. This resulted in the landscape we see today: brown/orange hills peaking at around 500 m, dappled with white caliche and dissected by valleys [**14**].

Ajaches' erosional stage was interrupted occasionally during the Pliocene Epoch (5.3 to 2.6 Ma) with eruptions of fluid lava around Las Breñas and southwards to Papagayo. The southern lava now occupies the south-eastern section of the Rubicón plain between Ajaches and the headlands of Los Coloradas and Punta del Papagayo. The mid-Pliocene, around 3 Ma, was a time of warm climate and high global sea levels so this gently sloping plain was periodically submerged, creating a distinctive marine platform at 30 to 40 m above modern sea level and sloping gently seawards [**17**].

16. Results of oxidation in soils

17. Papagayo 30 – 40 m marine platform

GEOSITES

The Geopark contains 85 Geosites: sites of special geological interest. Thirteen of these are submarine, leaving 72 on land, of which 53 are on the main island. They are categorised according to eight key properties: **palaeontology** (fossils); **hydrogeology** (water); **petrology** (rocks); **geomorphology** (landforms); **volcanism**; **tectonism** (earth movements); **sedimentology**; and **stratigraphy** (interpreting layers of rock). Some geosites are rather obscure and inaccessible, but many relate to accessible and photogenic landscapes or geological features. There are too many to summarise here but details can be found on the Geopark website and in Mateo et al (2019). (See going further).

18. Calcrete on Tinache

CALICHE

Although Lanzarote's dominant rock (basalt) is black, many of its land surfaces are white [**19**]: unusual for oceanic volcanic islands. This **caliche** is mainly calcium carbonate, forming crusts, layers and lumps of various thickness on (and in) soils, lapilli, ash and solid lava. Modern research shows that much of the calcium comes from wind-borne dust from the Sahara Desert. Some also comes from organic sands (fragmented shells) blown onto the island from the surrounding sea bed during periods of low sea level. A small proportion of the calcium is drawn up from the underlying basalt.

Much (but not all) of the caliche was formed during the Pleistocene Epoch [**5**] which, in the tropics, was a period of alternating wet and arid phases (the Sahara went through numerous "Green Sahara Phases"). North African dust was deposited across Lanzarote during dry phases, followed by downward washing of calcium carbonate during wet phases: hence the caliche. Older surfaces tend to have thicker and more extensive accumulations [**19**] whereas younger lava flows (especially those of the 1730's) have little or none. When it forms crusts on the land surface it's called **calcrete**, which gives many of Lanzarote's tephra cones some protection from weathering and erosion, admirably shown at Montaña Tinache [**18**].

19. Caliche

LOS AJACHES - GENERAL LANDSCAPE FEATURES

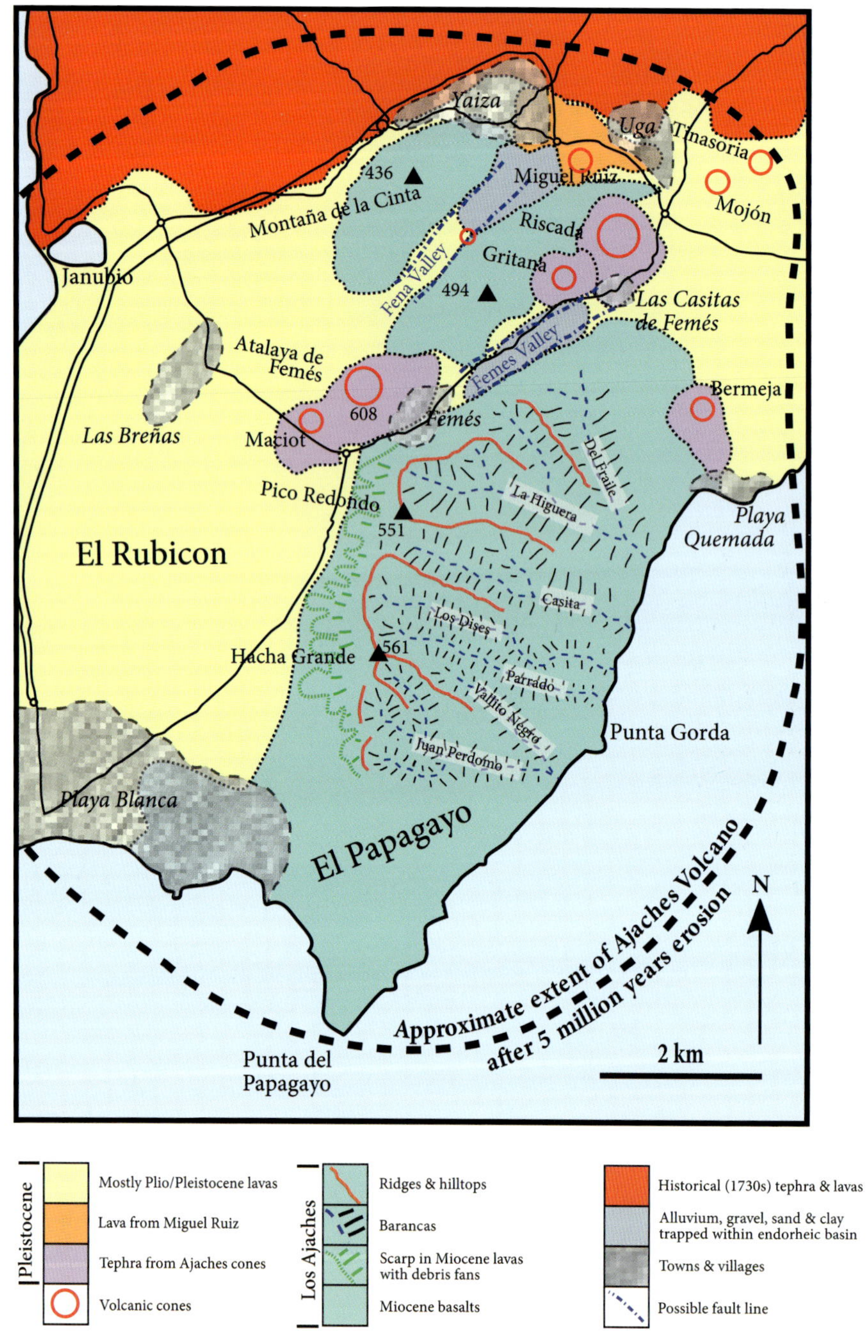

20. Los Ajaches landscape features

STROMBOLIAN ERUPTIONS

Named after Stromboli, an island volcano north of Sicily off southern Italy, these are mild but explosive gas-driven eruptions. Explosions last a few minutes and occur at irregular intervals, when the pressure has built up below the crater's magma pool. Existing rock fragments and molten lava droplets shoot upwards as glowing sprays before settling as a cone of tephra and/or spatter (accumulated lava blobs). Vertical gas/ash plumes range in altitude from 2 to 12 km. The immediate results are cinder **cones** (also called ash **cones**). On windy days far more particles accumulate on the down-wind side of the cone, so on Lanzarote the prevailing north-east Trade Winds have caused many of the island's cinder cones to have a low lip on the north-east side, shown here with the 2021 eruption on La Palma [**21**]. Most Strombolian eruptions also produce lava flows, especially in the closing stages when basaltic lava cuts a breach through the tephra crater wall or wells up in the crater and flows out over the lowest point. This sequence of (i) explosive eruption, (ii) formation of cone and crater, and (iii) lava flow is a common sequence for Strombolian eruptions globally. Most of Lanzarote's Pleistocene and historical cinder cones were produced by these single-event (**monogenetic**) Strombolian eruptions lasting between a few days and several months. Many of Lanzarote's volcanoes formed close to sea level, so the magma often became contaminated with shallow groundwater or sea-water. This generated eruptions that are more explosive than those on the other Canary islands, with intermittent bouts of Surtseyan activity, forming cones rather wider and lower than normal (see ***Hydromagmatic eruptions*** info box).

21. Effect of wind on Strombolian eruption, La Palma 2021

22. Columnar jointing

It also meant that lava occasionally emerged onto the sea floor where it cooled extremely rapidly to accumulate as small domes and bulges: **pillow lavas**, visible today in the Papagayo cliffs. Pliocene beach deposits are also extensive here (see Section 5: SEDIMENT).

2.1.3. Stage 3: Rejuvenation

Ajaches' third stage involved **rejuvenation** ("becoming young again") early in the Pleistocene Epoch when magma pierced the older rocks to produce a line of new volcanoes. The cones resulted from explosive **Strombolian** eruptions which generated tephra and lava: see ***Strombolian eruptions*** info box. One consequence of this rejuvenation was the blocking of the Femés and Fena valley outlets by new lava flows, so they became **endorheic** valleys (having no outlet). Deep fractures influenced the locations of these Pleistocene eruptions within Ajaches, notably Atalaya de Femés to the south-west (the highest peak in Los Ajaches, at 608 m) and Caldera Riscada to the north-east [**20**]. Section 3 (VOLCANIC REJUVENATION) covers these Pleistocene events for the whole island.

The highest peak in Ajaches' Miocene rocks is Hacha Grande in the south, at 561 m. For the casual observer, some of these hills in Miocene lavas may appear to be volcano-shaped: but they are quite definitely not volcanoes. They are merely remnant hills of almost random shape carved in the old volcanic rocks.

2.2. The north: Famara Volcano

Los Ajaches was into its erosional stages when eruptions started in the north, around 10 Ma: Famara Volcano rose above the waves. Lava and tephra emerged from numerous cones along a NE/SW linear fracture zone, producing a ridge volcano, constructed in **three distinct eruptive phases** each separated by long quiet spells. The **first phase** lasted from 10 to 8 Ma when Hawaiian eruptions produced

23. Famara Riscos

24. Famara plateau

25. Valle de Temisa

huge quantities of basaltic lava, each lava flow typically 1 – 2 m thick. Some were thicker, allowing hexagonal shrinkage cracks to develop in the top few centimetres as the lava became stationary and cooled. These cracks then progressed downwards as cooling and solidification proceeded, resulting in columnar structures, as here in the Famara cliffs [**22**]. Bending and other distortions of the columns resulted if the hot lava was still slightly mobile.

This was followed by about 1.5 million years of relative quiet, when an erosional landscape became established across this ridge island. Famara's **second eruptive phase** ran from 6.5 to 5 Ma and involved explosive eruptions creating pyroclastic flows and many tephra cones. The rocks can be seen in the upper parts of the spectacular Famara Crags (El Riscos) [**23**].

A further quiet period of around 1 million years was followed by a brief **third eruptive phase** between 3.9 and 3.8 Ma. Lavas dominated, similar to the first phase. They were draped across most slopes, making the valleys shallow and broad. Today these lavas form the higher parts of the Famara plateau, seen here looking north along the Guinate Valley towards Corona [**24**]. This was followed by a long quiet period when erosion dominated, from 3.8 Ma to 20,000 **years Before Present** (BP), ending with the eruption of the Monte Corona volcano cluster.

Tectonic movements (uplift, subsidence, tilting) and erosion occurred during the time gaps between each of Famara's eruptive phases, so the 3 sets of lava flows and tephra deposits do not rest parallel with each other, as can be seen in the cliffs overlooking Órzola.

Spasmodic volcanism continued through the Pleistocene Epoch, ending with the important Corona event of 20,000 BP. Now dominated by erosion, Famara plateau has been transformed into a brown-tinted landscape of erosion, with caliche-mantled surfaces dissected by steep valleys such as Valle de Temisa [**25**]. This valley records the fluctuating climate between humid and semi-arid conditions. It has an underlying V-shaped cross section, created by river erosion during wetter periods, but this was later smoothed into an open U-shape during more arid times when sediment from the valley sides (colluvium) accumulated in the valley bottom in the absence of a river.

The fault-guided and larger Barranco de Teneguime retains its V-shaped cross-section, a deep ravine cut by water into the Famara volcanics , but occupied by an early Pleistocene "intracanyon" lava flow [**37**]. The ravine sides provide good exposures of columnar jointing and many other features of the Famara basalts [**26**].

26. Barranco de Teneguime

LANZAROTE - GENERALISED VOLCANIC EVOLUTION

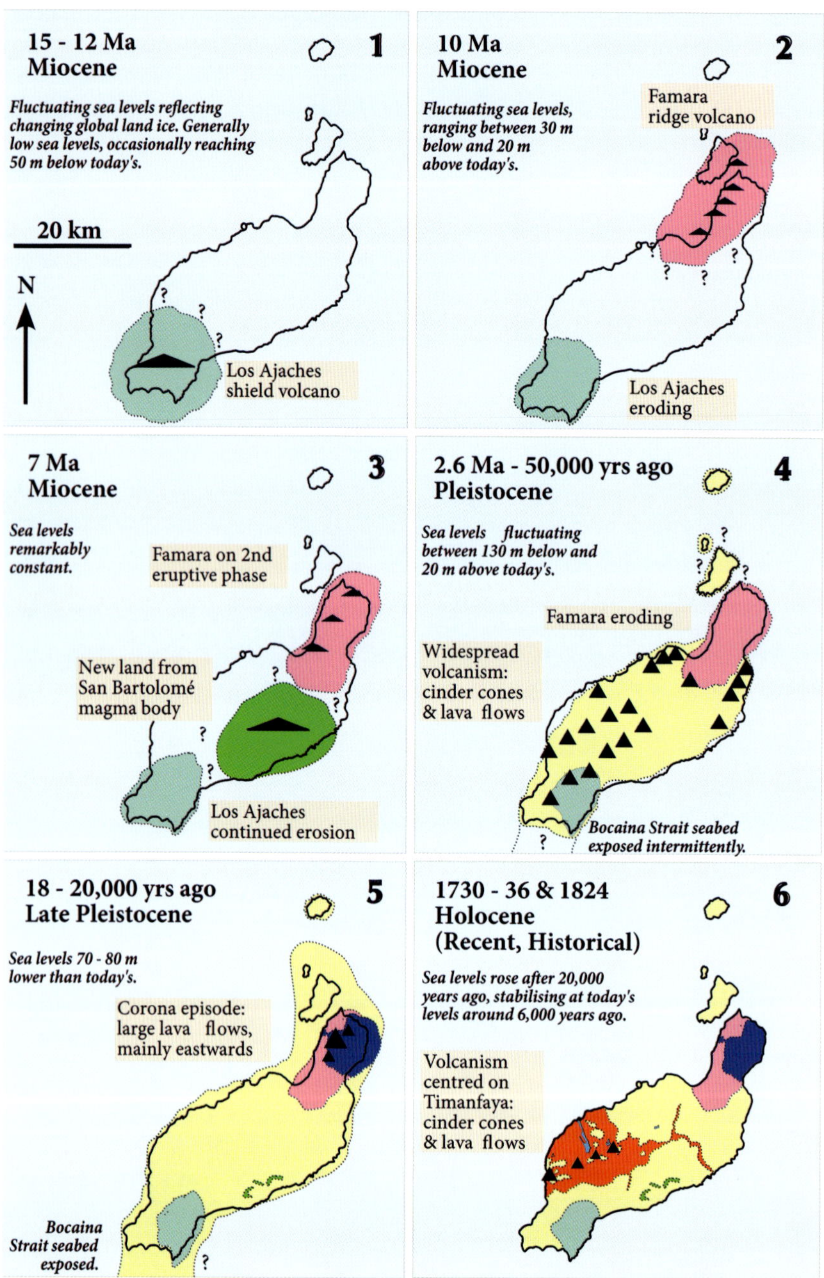

27. Lanzarote geological evolution

The Famara highlands peak at Peñas del Chache, the highest point on the island, at 670 m: see Panorama 2. Until 2 Ma the Famara plateau dropped away on the north, east and west as cliffs were created by landslides of various sizes and speeds. This plateau of eroded Miocene lavas was later covered on the south-eastern and southern edges by lavas in Pleistocene times, ending around 20,000 BP when the north-eastern side was covered by lavas from the impressive Corona Volcano and its neighbours.

2.3. The middle: San Bartolomé Volcano

Lanzarote's two islands probably became linked in late Miocene times by a third shield volcano centred around San Bartolomé. This was probably fed by a large, high density, intrusive igneous body about 10 km deep and 15 km across known to reside below the town. It may even have fed magma to the southern part of the Famara complex [**27**, map 3]. The original San Bartolomé volcanic massif is now largely concealed by younger lavas and cones, with just a few of its Miocene rocks exposed and quarried around Tías and Janubio.

At Janubio, lava flows of various ages and uncertain sources are interbedded with marine fossiliferous sedimentary rocks formed around 9.6 Ma (Miocene) when the platform was submerged by a shallow sea. For those with an expert eye, these complex relationships can be seen in the low cliffs on the south side of the Janubio lagoon, cliffs [**28**] which also show marine terraces cut by changing sea levels from late Miocene through to Pleistocene times. Ongoing research is attempting to unravel these complex relationships.

28. Cliffs, south side of Janubio lagoon

dollar

3. VOLCANIC REJUVENATION

3.1. The Big Picture

A lot happened during the Pleistocene Epoch. Global climate change produced an ice age, with ice sheets advancing and retreating across higher altitudes and latitudes as **glacials** and **interglacials** respectively. A glacial advance locks up water as ice on land, leaving less water to fill the ocean basins: so we get low global sea levels. Warmer interglacial phases cause the opposite: less (or zero) land ice, so more ocean water, so higher sea levels. Although Lanzarote had no ice sheets, being too low and too close to the Equator, it was certainly affected by these global events.

By the start of the Pleistocene Epoch, around 2.6 Ma, Lanzarote was the lowest and driest of all the Canaries, as it remains to this day. Then the hot spot got hot again, for reasons not yet fully understood. Starting in the late Pliocene Epoch, intermittent volcanic activity produced new volcanoes, lava flows and tephra deposits across the island, piercing the old eroded Miocene landscapes. Several valleys on the south east flanks of the Famara massif were occupied by "intracanyon" lava flows. The new eruptions of this **volcanic rejuvenation** produced Lanzarote's present-day general shape, modified only slightly by later sea level changes and the 1730's eruptions. Over 230 Pleistocene vents appeared on the main island and most of the volcanoes seen by visitors leaving the airport are of this age [**44**]. The airport is built on Pleistocene lavas [**30**]. Most of Lanzarote's Pleistocene volcanoes are cinder cones, formed by Strombolian eruptions.

Lanzarote's Pleistocene volcanic rejuvenation is puzzling: was the cause local or global? There was a world-wide burst of volcanic activity at this time, indicating some kind of global trigger. Some research suggests that Pleistocene's changing sea levels caused fluctuating pressure on the ocean floor rocks, with low sea levels (and therefore low pressure) allowing magma to ascend more readily. However, other research links this increased volcanism with high sea levels and higher pressure on the sea-floor: the opposite! And yet more research suggests it's caused by rapidly

29. (Previous spread) Caldera Riscada

◀ 30. Lanzarote Airport is built on Pleistocene lavas

changing sea levels in either direction! This is all very speculative and it might have been simply a new magma blob rising from the corridor below the island.

We can think of Pliocene-Pleistocene-Holocene times as a single cluster of rejuvenating volcanic events for Lanzarote, including Monte Corona and its neighbours around 20,000 BP. The Eighteenth Century eruptions were merely the latest batch of this ongoing rejuvenation, the only difference being that they were observed by humans. The Pleistocene Epoch is subdivided into four **ages**, but that's too much detail for us. We'll use the labels **early** and **late** informally, with the break around 780,000 BP, although dating of these eruptions is not straightforward and future research will undoubtedly produce different dates: see ***Rock and event dating*** info box.

3.2. Early Pleistocene eruptions

Early Pleistocene events usually involved complex, single-vent Strombolian eruptions of tephra and lava on the flanks of the two main uplands, producing large composite (ash/lava) or cinder cones, lava flows and tephra deposits around the two old plateaux, often burying their original boundaries [**13**].

3.2.1. Montaña Roja

Located on the southern flank of the much-eroded Ajaches massif, Montaña Roja appeared in late Pliocene or early Pleistocene times (dates conflict). Either way, it is one of the oldest volcanoes of Lanzarote's rejuvenation phase. It's a large composite cone [**31**], with lava flowing in all directions, especially to the north-east where it covered a wide marine platform previously cut across the Ajaches Miocene rocks: the Rubicón: see Panorama 8. Sea level was some 30 – 50 m above today's (it was an interglacial) so Roja lava often erupted on to a shallow sea bed and became interbedded with marine sandstones and limestones. The cone can be explored on the north-eastern side via rough paths which lead to the crater rim. There's a good view from the top.

31. Montaña Roja.

ROCK AND EVENT DATING

How do we know the actual dates of Lanzarote's volcanic events? One answer involves relative dating: the age of one rock relative to others. Usually a lava flow is older than the one above it but younger than the one below it for obvious reasons. But what about their actual (absolute) ages?

Rocks can be dated in many ways, ranging from historical records to sophisticated methods involving atomic analysis. **Radiometric dating** methods use radioactive elements in minerals which change (decay) to other elements at known rates. It is a topic beyond the scope of this book. Suffice to say that Lanzarote basalts older than 100,000 years can be dated using a radioactive form (isotope) of potassium (K) which decays to an isotope of argon (Ar): the K-Ar method. Another technique using two Argon isotopes, ^{40}Ar and ^{39}Ar, is now being used on Lanzarote rocks. Younger rocks can be dated using **radiocarbon** methods if they contain carbon or by **amino acid** methods if they contain organic remains such as shell fragments.

Another technique is **luminescence**, used on sediments, soils and tephra exposed to sunlight or heat from lava flows during formation. It uses light and other radiation absorbed by certain minerals during rock formation, notably quartz and feldspars. This energy is excited in the lab (using light or heat) and the emitted light energy (hence **luminescence**) is measured. The more luminescence the older the rock. Luminescence was used to date the sands and soils interbedded with Guatiza's lava flows at Mala, indirectly dating the lava at the same time, giving an age of 170,000 BP for the whole sequence.

It can work the other way too. For example, basalts at Janubio were recently dated at 9.6 Ma by using ^{40}Ar-^{39}Ar, indirectly dating the interbedded fossiliferous marine sedimentary rocks. So, in this case, igneous rock dating methods were used to date climatic records held in sediments.

Rocks don't date themselves! It takes research time, money and ingenuity, so plenty of Lanzarote's rocks and volcanoes have never been dated accurately by anyone. Approximate dating can be done by geological mapping, but there are various discrepancies and conflicting dates. Different researchers come up with different dates for the same volcanic event: Montaña Roja in Playa Blanca, for example, is dated variously between late Pliocene (around 3 Ma) and mid-Pleistocene (around 700,000 BP). Usually the most recent date is the most accurate – but not always.

3.2.2. Guanapay

Overlooking Teguise on the southern edge of the Famara uplands, Guanapay is an impressive early Pleistocene volcano [**33**], providing a great view from the top: see Panorama 3.

Guanapay is the largest of a cluster – the Teguise Group – which includes Montaña de Chimia north of the town. Most are cinder cones produced by Strombolian eruptions. The extensive lava flows from these volcanoes flowed mainly towards the east and south-east, covering the heavily-eroded Famara Miocene landscape and reaching the coast around Costa Teguise.

But lava also flowed northwards towards the coast, creating new lowland at the southern end of the Famara Cliffs [**32**], evidence that the Famara cliffs were created before these early Pleistocene times. These lavas blocked a pre-existing valley a few kilometres from Teguise, creating an endorheic basin, Vega de San Jose, seen here from the top of Guanapay [**36**]. The blockage allowed sediment accumulation to produce a broad fertile plain highly suited to agriculture.

32. New land formed beneath Risco de Famara in Pleistocene times

33. Guanapay

3.2.3. Riscada, Gritana, Bermeja & Miguel Ruiz

Several large volcanoes emerged at this time around Los Ajaches uplands, during its rejuvenation phase. At the north-eastern end of the Femés valley Caldera Riscada [**29**], Caldera Gritana [**35**] and several smaller volcanoes pierced through the eroded Ajaches rocks, creating cinder cones and spreading lava to the north and east, around Yaiza and Uga. These flows blocked the north-eastward sloping Femés valley, creating an endorheic basin from which valuable brown soil is extracted for agricultural use [**16**]. Miguel Ruiz lava had the same effect in blocking the Fena valley. Montaña Bermeja, overlooking Playa Quemada, was also formed around this time [**34**].

34. Bermeja, near Playa Quemada

35. Caldera Gritana, showing lip on windward side

36. Vega de San José

LANZAROTE: GEOLOGICAL LANDSCAPES

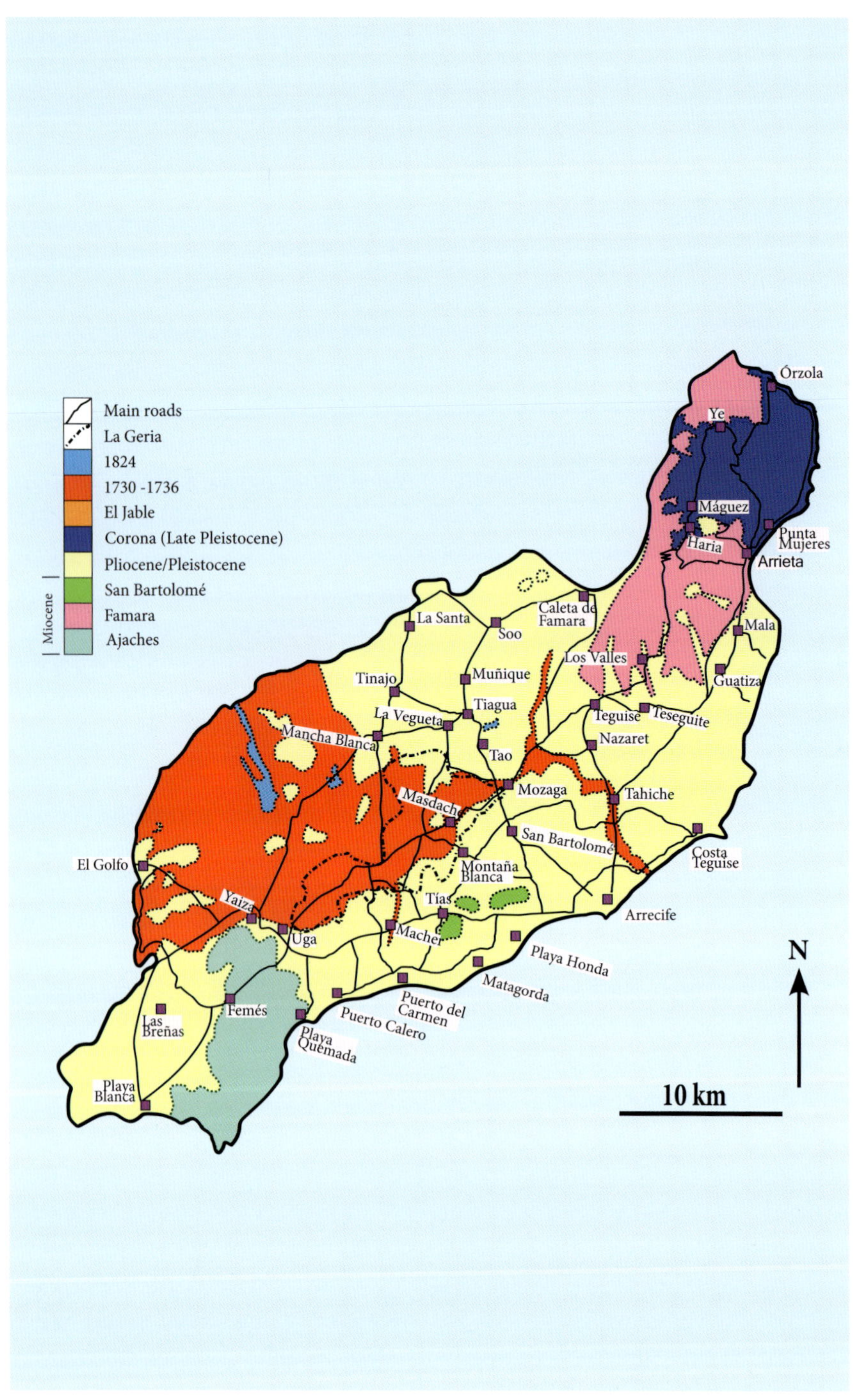

37. Lanzarote geological landscapes

LANZAROTE: MAIN VOLCANOES

38. Main volcanoes. See map of central zone on page 91

3.2.4. Atalaya de Femés

This large Pleistocene volcano [**39**] erupted around 780,000 BP, piercing the Ajaches massif and spreading lava to north, west and south [**37**]. It produced a large tephra cone and much basaltic lava. To the north its lava flowed down the Fena valley towards Yaiza. To the south-west, across the Rubicón platform, its lavas sit on the Miocene 30 m marine terrace, cut into the Miocene lavas (possibly from San Bartolomé Volcano). In places, Atalaya lavas rest on older Roja lavas, both sets being interbedded with marine sediments. Atalaya lies at the south-western end of the 45 km long NE/SW-orientated fracture zone which became the focus for later Pleistocene volcanoes [**43**].

39. Atalaya de Femés

VOLCANO NAMES

Warning! Volcano names have evolved over the centuries, sometimes resulting in several names for the same cone or the same name for several different cones. For example, there are half a dozen Bermejas, quite a few Blancas and several Mojón, Rodeos, Rojas and Quemadas [**38**]. Santa Catalina has been used for at least 3 different cones and Rilla used for Santa Catalina and Corazoncillo. This is all confusing but understandable. Even official (government) names are not that helpful because they might not be used in maps, texts and websites used by the general public. However, its not always a problem for us because some cones are small and geologically less significant. We just have to be careful and consistent. The Caldera/Montaña distinction sometimes helps. For example, Caldera Blanca is the huge cone north of Timanfaya but Montaña Blanca is the cone overlooking the airport and next to the village with the same name.

Geologically the term **caldera** describes an extremely wide crater formed by large-scale collapse of a volcano into its underlying magma chamber: but there are no true calderas on Lanzarote. The label is used on the island for a volcano that is somewhat broader than normal, reflecting its explosive eruptions.

3.2.5. Caldera de Maciot

This amazing small tephra cone sits on the south-west flanks of Atalaya [**39**] and shows a marked colour change from black to reddish brown [**40**]. The black tephra (mainly lapilli) was altered in the ***late stages*** of volcanism by extremely hot water (***hydrothermal***) circulating through the permeable ash and lapilli. This caused the iron content of the basalt minerals to oxidise rapidly, ***altering*** the colour from black to shades of red, orange and brown. Understandably, this is called ***late stage hydrothermal alteration*** and is common across the island and, indeed, the globe. Both black and brown tephra is extracted for use throughout the island as an agricultural mulch and for decorative purposes [**41**].

40. Caldera de Maciot

41. Maciot quarry

3.3. Later Pleistocene eruptions

Most volcanoes seen by Lanzarote visitors were created in later Pleistocene times, from around 780,000 to 11,000 BP, and were responsible for creating a large proportion of today's island: see ***Volcano Names*** info box and website in GOING FURTHER. Lava and tephra covered the entire central part of the island and much beyond: see Panorama 7. At least 9 distinct NE/SW linear volcano clusters emerged [**43**], the largest running from Montaña del Mojón near Uga to Montaña de Tahíche. This 22 km range comprises 16 substantial cinder cones [**44**], the highest being Montaña Guardilama at 603 m [**42**]. Towards the eastern end, between Tahíche and San Bartolomé, sits Montaña de Zonzamas, the island's main landfill site [**47**]. Other linear clusters include the Guatiza chain in the north-east [**66**], the Soo chain [**46**] and the string of volcanoes later surrounded by the 1730's lava flows of Timanfaya, with El Golfo at its western end [**54**].

42. Montaña Guardilama

LANZAROTE: MAIN FISSURES AND FRACTURE ZONES

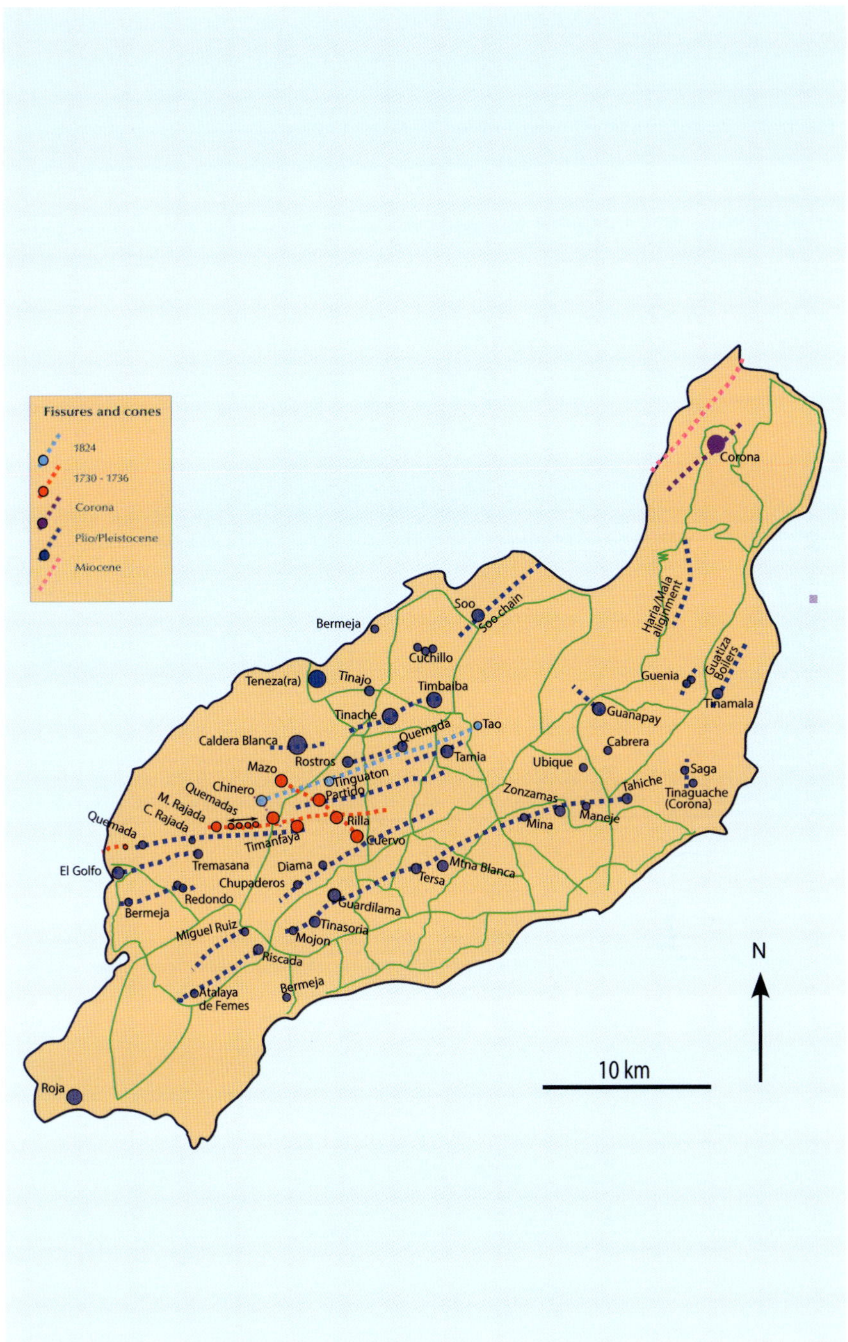

43. Volcano alignments. A generalised map from various sources, with selected volcanoes.

44. Pleistocene cones near airport

45. Montaña de Tinasoria

46. Soo chain

HYDROMAGMATIC ERUPTIONS

Hydromagmatic eruptions are explosive because the magma is mixed with water. There are two main types: **Surtseyan** and **phreatomagmatic**. Surtsey erupted in 1963 off the Icelandic coast. It was highly explosive because magma mixed with seawater as it emerged, producing a Surtseyan eruption: intermittent, explosive, dangerous and very loud! Surtseyan eruptions occur in seawater. When groundwater is involved we get **phreatomagmatic** eruptions (the phreatic zone is the saturated rock below the water table). The result is tephra on a grand scale, often with no lava at all. Such explosive eruptions are particularly plentiful on Lanzarote compared with the other islands because of its low altitude – the land surface is close to the water table. Many of Lanzarote's volcanoes began their eruptions as hydromagmatic types but ended as Strombolian, producing wide, low cones with variable proportions of lava. El Golfo [**54**] is Lanzarote's best and most accessible example of a volcano produced by a powerful Surtseyan eruption. El Cuchillo (Pleistocene) is a good example of a phreatomagmatic eruption. The 2011/12 eruption off El Hierro was Surtseyan.

47. Zonzomas, with Tamia middle distance right and Caldera Blanca far distance left

The Pleistocene volcano chain running through the 1730's lava fields can be confusing for the casual observer. Many of these cones and upland areas were later surrounded by the 1730's lava, leaving them as isolated **islotes**. Compared with the fresh, dark grey lava and tephra deposits of 1730's, the Pleistocene islotes are generally more rounded (eroded) and usually paler (more caliche, soil and vegetation). They often contain remnants of old stone walls and they turn green for a few weeks after heavy rainfall.

48. Lichen

Many of the Pleistocene cinder cones have typically horseshoe-shaped crater rims with a lip on the north-eastern side, formed by wind action on the airborne ash, as described in the ***Strombolian eruptions*** info box [**35**]. This low lip often became a breach through which lava later flowed, lowering it even further and steering the lava in that direction. This can be seen at Tinasoria, a large, late Pleistocene volcano 3 km west of Mácher where vine growing has extended into the crater [**45**].

49. El Golfo, wet tephra

3.3.1. Caldera Blanca, Risco Quebrado and Montaña Caldereta

Caldera Blanca and Risco Quebrado [**51**] comprise an overlapping pair of large ash and tuff cones formed as a result of **phreatomagmatic** eruptions, probably in mid-Pleistocene times: see ***Hydromagmatic eruptions*** info box. Access is via a track from Mancha Blanca which passes Montaña Caldereta. Caldera Blanca is one of the largest Pleistocene volcanoes, with a crater diameter of 1.1 km [**50**]. It looms over the northern part of the 1730's lava field as a large islote, allowing spectacular views in all directions from its crater rim. Blanca's white coloration is caused largely by lichen colonisation [**48**] rather than the more familiar caliche. Blanca/Quebrado and Caldereta are used for Mars-related research because they provide important data about possible Martian hydromagmatic processes.

50. Caldera Blanca crater, looking east with Teneza left distance

51. Caldera Blanca

3.3.2. Timanfaya's Pleistocene chain of islotes: El Golfo to Tamia

Several parallel Pleistocene volcano chains run through the Timanfaya region, each revealing a local weakness in the general NE/SW crustal fracture zone and each producing islotes surrounded by younger 1730's lava fields. El Golfo sits at the western end of the major Pleistocene chain, with Tamia 20 km to the east [**43**].

Located on the coast and with an unknown precise date, El Golfo erupted as a single event which probably lasted weeks or months: so it is monogenetic. This eruption was highly explosive because the magma mixed with seawater – a Surtseyan eruption: see ***Hydromagmatic eruptions*** info box. The cone was constructed of tephra of all sizes, although ash dominates because of the high energy explosive eruptions. No lava flows were produced. The abundance of sea-water caused much of the tephra to be erupted as wet jets which flowed rapidly down the volcano sides, depositing ash and lapilli in layers [**49**]. El Golfo was sliced (irregularly) down the middle by sideways explosive eruptions towards the end of its eruptive phase, removing its western half, so we can now conveniently view the volcano's interior [**54**]. Subsequent sea erosion has also helped to expose its insides. Unsurprisingly, El Golfo is a major tourist attraction.

When magma cools extremely rapidly, as it did here in contact with seawater, it often produces a **volcanic glass** rather than a normal crystalline basalt. Fragments of all shapes and sizes, many glassy and filled with gas bubbles, were hurled into the air to create tephra, with huge quantities becoming tephra flows (often wet) which built the Golfo cone. Some of the ash was blown by the wind to produce small dune systems on the volcano side, but particles became consolidated and welded (stuck together) to produce **tuff**, so this is actually a **tuff cone** rather than a cinder cone. Larger bombs and blocks up to 50 cm across got pushed out of the vent and tumbled down the volcano side. When they fell on fresh, soft wet tephra they often produced a sag structure.

You will see green **peridot** at El Golfo. It's a semi-precious gemstone found in abundance across Lanzarote and often used in jewellery. The scientific name of this mineral is olivine and the rock dominated by **olivine** is called **dunite**. Olivine is iron magnesium silicate (Fe,Mg SiO_4) which crystallises out of magma slowly (and so forms large crystals) and at pretty high temperatures (1,200 to 1,300^{O} C): so it is the earliest mineral to crystallise as magma cools, forming lumps of (solid) dunite up to 30 cm across in the (liquid) magma on its way up [**52**].

52. Dunite xenolith

53. El Golfo, palagonitisation

54. El Golfo

Originating in the upper mantle some 20 km down, these dunite **xenoliths** ("foreign rock") become embedded in the ascending basalt magma, many disintegrating during the explosive eruptions to form their own special ash or lapilli particles. The helium, neon and argon held within Lanzarote's xenoliths provide important information about Earth's mantle.

Another type of xenolith originated at shallower depths, in the limestones and sandstones making up the highest parts of the crust. Fragments were caught up in the rising magma and ended up as pale, frothy lumps embedded in the Pleistocene lavas, typically about 20 cm across. More recently such frothy "xeno-pumice" xenoliths were found floating in the sea after the 2011 eruption off El Hierro – and they contain tiny datable fossils (nannofossils) from the limestone which survived the intense magmatic heat, giving strong modern evidence for the plume theory. Xenoliths of various types are excellent messengers from the depths, revealing much about the composition and behaviour of not only the mantle itself but also the crust through which the magma ascends.

There's a lot of **palagonite** at El Golfo. This is a pale yellow/brown crumbly, clayey material produced when water reacts with tephra, volcanic glass or basaltic magma while it is still hot, or even molten. It's a complicated chemical process which we don't need here. **Palagonitisation** took place during El Golfo eruptions and also after the tephra had settled. The results can be seen in the seawards facing slopes behind the green lake as the pale yellow/brown zones adjacent to unchanged darker areas [**53**].

55. Bermeja, near El Golfo

56. Halcones

Three new small craters were formed on the northern side of the main Golfo cone, probably years after the main event. Unlike earlier eruptions, these produced some lava, although they were dominated by pyroclastics.

Several shorter NE/SW chains form islotes within the Timanfaya lava fields. One sits about 3 km to the south of Golfo, with Montaña Bermeja [**55**] at its western end (with car park adjacent). This chain extends eastwards for about 5 km, forming a particularly large upland islote, used by the military and traversed by the Yaiza-Golfo road: see Panorama 5. A few kilometres north of Golfo and entirely surrounded by 1730's lava, Montaña de Halcones [**56**] is the crescent-shaped remains of a large coastal volcano, cut-through by several large dykes. Another 6 km NE/SW Pleistocene chain emerged just to the north of the main Golfo/Tamia chain, a few km south of Mancha Blanca, with Rostros in the west and Montaña Quemada ("burned") in the east, near to Tamia. This short chain also includes a second Caldera Quemada.

57. Corazoncillo

Surrounded by 1730's lava flows, Islote de Hilario, Montaña Rodeos and Caldera del Corazoncillo [**57**] are central in the main Golfo/Tamia chain of Pleistocene cones, Hilario providing the base for the Timanfaya visitor facilities. Corazoncillo exhibits the broad, low profile so characteristic of Lanzarote volcanoes because of the frequent explosive hydromagmatic phases within the normal Strombolian eruptions. At around 500 m across, its crater is abnormally wide for a cone of this size. This stunning volcano, identified by some as a 1730's cone, can be seen (and photographed) from the dedicated coach within the Timanfaya protected tourist zone or from the public roadside. Rodeos and Corazoncillo sit just on the western side of the important Partido/Señalo complex formed in the 1730's. On the north-eastern side of Partido/Señalo the Pleistocene chain continues with Montañas Los Rodeos, Ortiz, Tizalaya, Meseta and Tamia.

58. Lava aa

59. El Cuchillo

3.3.3. El Cuchillo

Explosive hydromagmatic eruptions produced El Cuchillo. Its crater is 1.5 km across and is lowest on its northern side, allowing farmers to develop fields across the accessible crater floor. It is actually a cluster of 3 linked craters, with its neighbours Montaña Mosta [**61**] and Montaña Cavera produced in separate eruptive phases [**59**]. Exceptionally wide and low craters formed in this way are called **maars** and are often occupied by lakes in wetter parts of the world. Cuchillo's base is only 40 m above sea level, so the mixing of the rising magma with both saline and fresh groundwater is not surprising. Like El Golfo, the tephra/basalt cone has been palagonitised in places, giving the distinctive beige colouring. Another distinctive feature is the presence of **accretionary lapilli** [**60**], lapilli-sized particles (2 - 64 mm) formed by (smaller) ash particles sticking together, usually while they are still airborne.

They are typical of hydromagmatic eruptions, being formed when the particles are wet. The water initially holds them together, but minerals are later precipitated out from the acidic water, cementing the grains together and making the accretionary lapilli more robust. Sometimes they flatten when they land.

60. Accretionary lapilli

61. Montaña de Mosta

62. Old quarries, Tinamala

3.3.4. Guatiza Volcanoes

The small NE/SW 6 km chain of volcanoes around Mala and Guatiza in the north-east of Lanzarote has been dated (in 2012) at 170,000 BP [**66**], although some earlier authors dated it as 10,000 BP and others around 30,000 BP. Known as Los Calderetas de Guatiza (Guatiza Boilers), they form a ridge on the east side of the road linking the two villages, running from Montaña Cobrada (also called Mojón) in the north to Montaña Tinamala in the south, located near the filling station south of Guatiza. Basalt lava flowed northwards and eastwards from the northern cones but southwards and eastwards from the southern cones. Indicated by the bright brown colour of its ash, Tinamala cone underwent hydrothermal alteration immediately after formation and its ash has been extensively quarried on its western flanks [**62**]. Much of Tinamala's ash is consolidated, so it's actually a rock called tuff. On its southern slopes it has been quarried using a method which leaves distinctive vertical cliffs and horizontal benches where blocks were cut using a special machine devised by Jesus Soto in the 1950's [**65**].

63. Bread-crust bomb from Montaña Colorada

This volcano also produced **bread-crust bombs**, formed from airborne lava blocks when the outer solidified layer became cracked as gases within the bomb's molten interior expanded during transport. Some of these can be seen in the Cactus Garden at Guatiza [**63**].

The lava flows from Cobrada spread over the coastal plain east of Mala, burying, baking and preserving the underlying soils. In a much-studied sand pit near Mala the 3-fold sequence consists of baked soil overlain by a 1 m thick lava flow with younger sandy dune deposits on top [**64**]. The soils and dunes have been dated at 170,000 BP: see Section 5, SEDIMENT. A similar (luminescence) dating procedure was undertaken on a palaeosol (ancient preserved soil) interbedded between two lava flows on the coast at Los Cocoteros, east of Guatiza. These were previously thought to come from (much older) Guanapay, but the same date of 170,000 BP was found, confirming Guatiza as the source. It seems that these Guatiza Boilers were the penultimate eruptive phase in the Pleistocene Epoch which ended with the big eruptions of Corona around 20,000 BP.

64. Mala sand/palaeosol deposits

65. Quarry on south side of Tinamala

66. Guatiza chain

3.4. Monte Corona

Monte Corona is well-named: it means "crown" and this volcano certainly crowns the northern Famara uplands in a majestic way, seen here looking northwards across Máguez, with Los Helechos to its left: (see Panorama 1). Following the eruption of the Guatiza volcanoes in the north-east around 170,000 BP, a long quiet period ended with the eruption of a string of volcanoes on top of the old Famara massif, the three main ones being Corona (the largest), Los Helechos and La Quemada. Corona's lava flows have been dated at around 20,000 BP (i.e. Pleistocene), coinciding with the peak of the last glaciation when sea levels were about 70 m lower than today's, although it is likely that Helechos and Quemada erupted earlier, around 91,000 BP.

Most lava from all three cones flowed eastwards [**68**], although a small proportion of Helechos lava flowed westwards down a small valley close to Guinate, spilling over the Famara cliff [**67**], leaving clear evidence that the cliff itself pre-dated the eruption.

Corona lava is varied. When fluid basaltic lava cools it often produces a thin crust which gets crumpled and folded by the underlying hot molten lava. This "ropy" lava is pahoehoe, an Hawaiian name [**6**]: see ***Igneous rocks and lava*** info box. By contrast, viscous (sticky) lava produces an irregular blocky surface of jagged fragments. This rubbly (or blocky) lava gets carried forward by the underlying flow. This is called aa, another Hawaiian term [**58**].

67. Corona lava over Famara Riscos

FAMARA AND CORONA; SIMPLE GEOLOGY

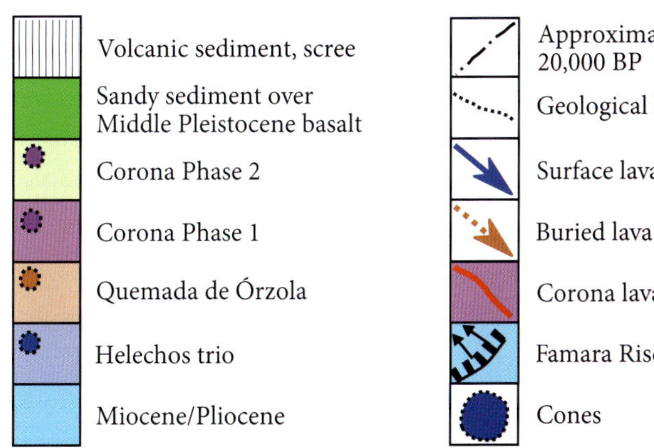

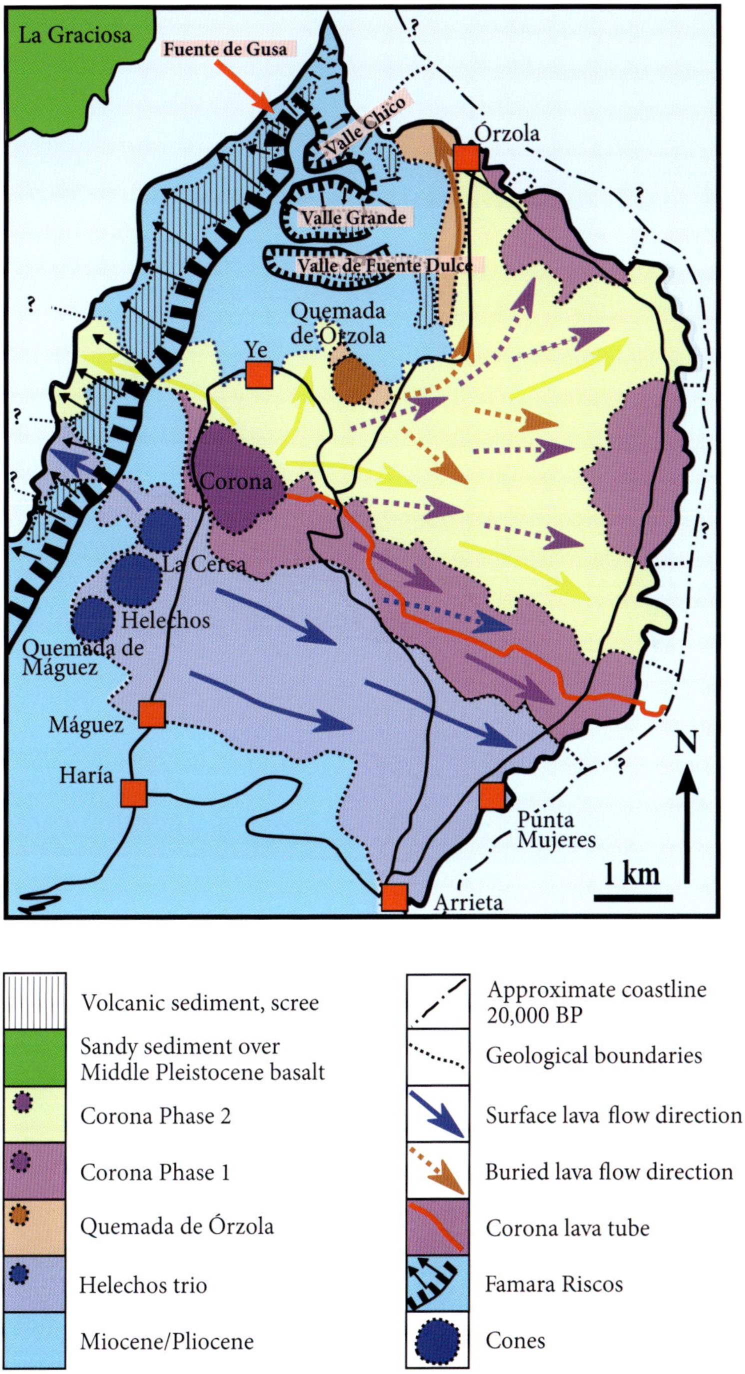

68. Famara and Corona, basic geology

69. Corona crater lip

70. Transported blocks, Peñas de Tao

71. Euphorbia balsamifera

Corona itself had three phases of eruption: (1) cone formation, (2) pahoehoe lava flows, (3) aa lava flows [**68**]. Phase 1 involved Strombolian eruptions of ash and lapilli which built the main cone, relatively higher and steeper than most Lanzarote volcanoes because of the vast quantity of tephra. The huge volume of Phase 2 fluid pahoehoe lava then ran south-eastwards direct to the sea, flowing between the earlier lavas of Quemada de Órzola to the north and Helechos to the south, extending the coastline by 100 m.

Phase 3 of Corona's eruptions comprised large blocky aa lava flows to the east and north-east, covering most of Quemada's older flows and forming an extensive area. Blocks of various sizes, some as big as houses and over 30 m across, were rafted on these sticky lava flows [**70**], the area now known as Peñas de Tao. The largest blocks were formed during collapses of the crater rim [**69**], allowing lava to spill out of the crater and flow rapidly down-slope. Some blocks increased in size by accretion as molten lava stuck on and solidified as everything flowed downslope. Today this blocky lava forms the spectacular **Malpais** landscape ('bad country'), vegetated by distinctive Euphorbia species and forming most of the coastline between Jameos del Agua and Órzola [**71**]. One of Corona's major lava flows reached the sea and produced a sharp headland, Punta Escamas, 2 km north of Jameos del Agua. Explosive eruptions occurred when the lava hit the seawater, generating distinctive mini-cones and craters.

During its last eruptive phase, Corona sent a smaller and more fluid lava flow north-westwards over the Famara cliff just to the north of the older tumbling flow from Helechos.

FEATURES OF THE MONTE CORONA LAVA TUBE

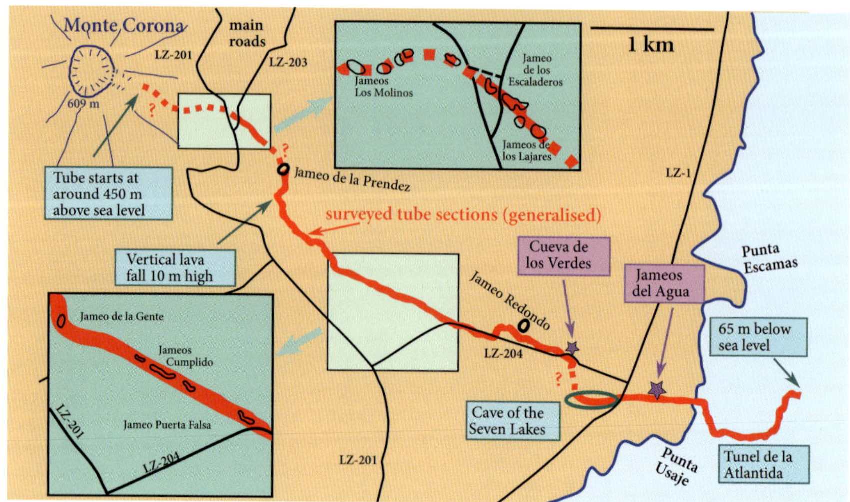

72. Corona lava tube

Lanzarote has one of the best lava tubes in the world, formed in one of Corona's Phase 2 fluid lava flows which ran eastwards towards the sea: see ***Lava tubes*** info box. And it's one of the longest, widest and most accessible too. It runs from the lower slopes of Monte Corona (at an altitude of 310 m) eastwards to the sea near Jameos del Agua: and then continues beyond the modern coastline, falling a total of 374 m. At 10 km overall (13th longest in the world) it is complex, with unknown or blocked sections, so it is not quite continuous [**72**]. In places its diameter is over 30 m, among the largest in the world. In some places it has several levels, one conduit running above another and several lava falls have been discovered. It is on public view at Jameos del Agua [**74**] and Cueva de Los Verdes where the lighting by Jesus Soto produces stunning and surprising effects [**73**]. Located between Agua and Verdes but closed to the general public, the Cave of Seven Lakes is much favoured by cavers [**75**].

The last 1.7 km stretch of Corona's lava tube descends to 65 m below present-day sea level, so it is flooded, giving the longest submarine lava tube in the world: Túnel de la Atlántida. During exceptionally high tides the lake in Jameos del Agua overflows and prevents visitors passing freely on the path alongside. Lava tubes cannot form below sea level because lava solidifies rapidly when it hits water, causing the tubes to become blocked. Clearly, therefore, this lava tube must have formed when sea levels were at least 65 m lower than today's and the coastline around 2 km further to the east, in fact the time of the last glacial maximum at 20,000 BP. It all ties in with the Pleistocene global events outlined above. Simple really.

73. Cueva de Los Verdes

74. Los Jameos del Agua

75. Cave of Seven Lakes

LAVA TUBES

Lava tubes are shallow tunnels within lava flows, ranging in diameter from less than one metre to over 30 metres, which form in fluid lava when it flows down gradient and starts to cool. The initial formation of a solid surface crust insulates the underlying lava which remains hot and mobile. The flow may become concentrated into subsurface tubes, either by chance or by following the underlying topography. Because of the insulation, fluid lava is able to travel great distances beneath the crust as tubes, flowing and branching in all directions, including up-gradient, at various levels and often creating a network of linked passages. Once a tunnel is formed it becomes a focus for more mobile lava, so it tends to stay hot and perpetuate itself until the supply of lava from the volcano gets cut off. The lava tunnel then drains, leaving a lava tube. In places the roof may collapse, creating holes which are called **jameos** on Lanzarote.

Lava tubes occur across Lanzarote in lavas of all ages, with a total length approaching 50 km. Many are named, one is inhabited and all attract cavers from around the world. Tubes within Timanfaya National Park are not well-known, but there's one which is 11.7 km long (Sistaqma Sin Nombre - Cueva de los Pescadores), making it the 10th longest in the world. Many contain **speleothems** ("cave deposits") of gypsum, a hydrous calcium sulphate ($CaSO_4.2H_2O$). This has many forms: thick white powder or crystalline crusts on cave floors; crusts and lumps across cave walls and ceilings; stalactites suspended from the ceilings; filling various cracks and gaps in the surrounding rock; and crystals of many types. Research at Cueva del Covon [**76**] shows that the calcium originates as seawater spray which enters the cave directly via air currents or indirectly by seeping downwards through the rock.

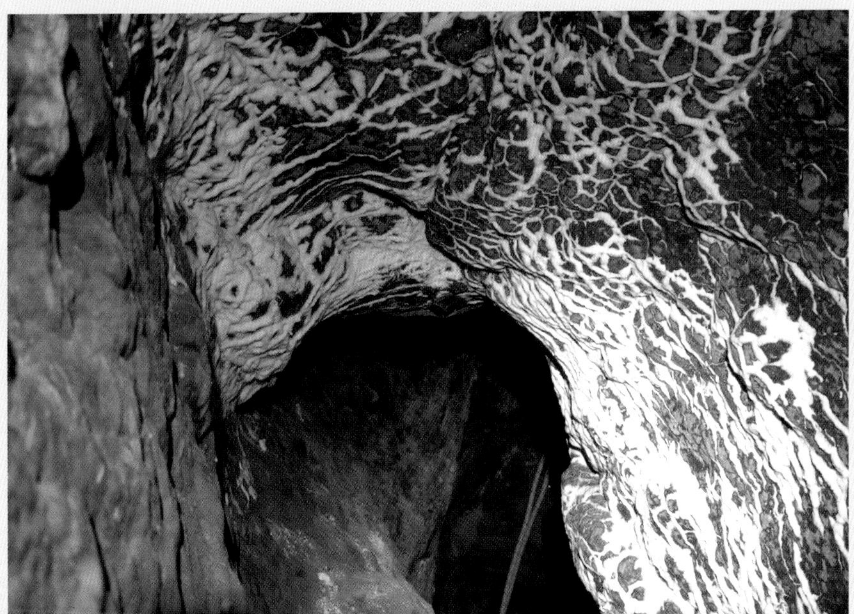

76. Gypsum in lava tube

4. HISTORICAL EVENTS

4.1. Overview

Centred on Timanfaya, Lanzarote's volcanic events of 1730 – 36 were spectacular, the world's third largest basalt eruption (by volume) of the last 1,000 years. Around 300 vents of all sizes produced up to 5 km^3 of lava and tephra, with further submarine eruptions offshore. In places the lavas reached 100 m in thickness. Eruptions were mainly Strombolian and Hawaiian, with significant explosive phases, some recent research suggesting that explosive activity was much greater than previously thought. Because of its scale and diversity of geological features and international importance, Timanfaya is one of Spain's Global Geosites.

Nearly a quarter of the island (around 225 km^2) was covered with lava or tephra, including much of the fertile land, together with 8 small towns and 14 villages. Over 700 houses were buried, along with 3 chapels, a granary, 1,500 water cisterns, several large cattle meadows and numerous tracks and paths. In many places the lavas flowed around the higher parts of the Pleistocene landscape, leaving **islotes** of old volcanic cones (see Panorama 5) and abandoned farmland, isolating buildings and field boundaries, as seen between El Golfo and Yaiza. Many settlement names survive today as volcano names, such as Santa Catalina, Chimanfaya (now Timanfaya) and Rodeo [**80**].

Eruptions occurred in five phases and lasted nearly 6 years, the volcanic gases producing acid rain which contaminated the island's soils, leading to famine and large-scale emigration to the other islands. Of the island's 4,967 population, about 2,000 were directly affected by the eruptions through loss of buildings and farmland, although only one life was lost. The rest were affected indirectly in many ways and northern Europe felt the impact too, the 1732/33 winter being exceptionally cool, dull and foggy with a prolonged airflow from the south.

77. (Previous spread) Rafted block from Cuervo, with Señalo as backdrop

◀ 78. Calderas Quemada looking east towards Timanfaya massif

4.2. Timanfaya's five phases

Each of Timanfaya's five eruptive phases, first articulated in detail in 1990 by JC Carracedo and others, comprised intermittent eruptions from clusters of volcanoes separated by quiet periods of several weeks or months. Most eruptions were Strombolian, producing monogenetic cinder cones and ending with lava flows. A typical single cone eruption lasted a few weeks or months, the longest ones over a year. Most eruptions occurred along a 14 km line of crustal weakness orientated roughly NE/SW, although the first four were on a NW/SE fissure at the eastern end of the main fissure [**43**]. About 30 large volcanic cones were formed over the 6 years, clustered into ten distinct eruptive centres.

Various records were kept at the time, notably by the priest of Yaiza, Don Curbelo, but only until December 1731 when he left the island. He returned in May 1732 but didn't finish his diary notes until 1744. Modern research suggests he may have got some dates wrong, recording 18 October 1730 instead of 10 October and recording 20 January 1731 incorrectly as 10 January. The island council (Cabildo), the Canary Islands cathedral and the Canary Islands Royal Court of Justice also kept records, although there are contradictions across all sources, so we won't dwell on them too much.

79. Montaña de Las Lapas o del Cuervo

LANZAROTE: MAIN VOLCANOES, CENTRAL AREA

1824
1730-1736
Pleistocene
Settlements
Timanfaya National Park points
Roads
Restricted roads

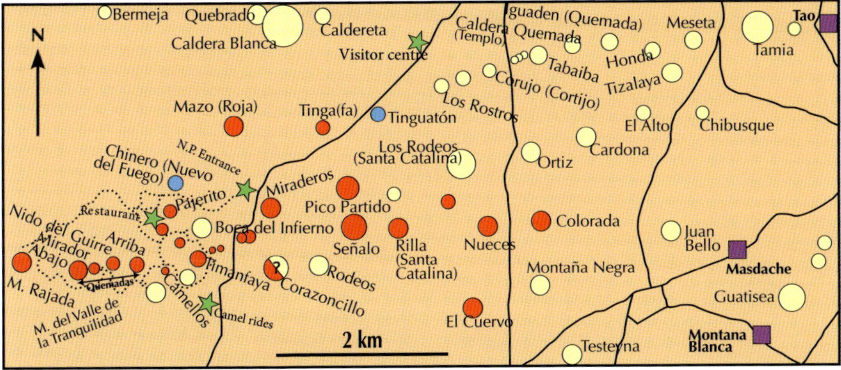

80. Main volcanoes - central area

81. Caldera Blanca/Caldereta bottleneck

82. Partido lava channel

4.2.1. Phase 1: Montaña del Cuervo, Pico Partido, Caldera de la Rilla (Santa Catalina) and Mazo

After several months of seismic tremors, on 1 September 1730, just after 9 pm, a large earthquake was followed by big, loud volcanic eruptions which persisted for 19 days as new vents were opened up. Curbelo named the Timanfaya district as the location, the first volcano now known as Montaña de los Lapas o del Cuervo, also called Caldera de los Cuervos. [**79**]. This sits at the south-east end of the short NW/SE fracture zone mentioned above [**43**]. Events followed the usual Strombolian pattern: a spatter cone formed over a few days followed by fluid lava which flowed northwards [**86**], destroying the villages of Maretas, Santa Catalina and Mazo (on 11 September). Curbelo recorded a particularly loud eruption on 7 September when a large chunk of Cuervo's crater edge was forced northwards from the cone, rafted on the fluid lava and coming to rest 150 m away [**77**]. It diverted the subsequent lava from a northward to a north-westward direction. Cuervo's fluid lava eventually reached the coast and killed many fish. The inside of Cuervo is readily accessible via a footpath from a roadside car park, with interpretation boards and well-marked routes.

83. Surface features, Timanfaya

Phase 1 ran from September 1730 to March 1731, with four main eruptions on the initial NW/SE fissure: Cuervo from 1 September, followed by Rilla (Santa Catalina) and Pico Partido from 10 October and then Mazo from 20 January. Rilla is accessible on foot from the LZ-56 road and has a spectacular crater lava lake which domed upwards during its final cooling [**85**]. The volcano was named after the village (Santa Catalina) buried by the lava, its inhabitants *moving en masse* to Los Valles in the north. Partido's lava lake accumulated during its active phase and was drained by a lava channel which shows spectacular overbank accumulations [**82**]. Partido also has a stunning lava tube, 170 m long and running 25 metres underneath the famous lava channel, often visited by cavers [**84**]. These Phase 1 lavas were very fluid and generally flowed northwards, round the large Pleistocene volcano of Caldera Blanca creating a steep gradient through the bottleneck between Blanca and Caldereta [**81**].

84. Partido lava tube

85. La Rilla crater

86. Cuervo, from across its northern lava field

Situated at the north-western end of the NW/SE fracture 1 km south of Caldera Blanca, Montaña Mazo (also called Negra and Roja at various times) is controversial: see Panorama 6. It was mapped in 2004 by IGME (Spanish Geological Survey) as Middle Pleistocene (around 600,000 BP), but others date it somewhere between 1732 and 1736 (Phase 4, see below): a big difference! However, detailed fieldwork and forensic study of several documents in 2020 tell a detailed story and indicate yet another date. Summarised briefly, the debris to the north, east and west of the cone is exceptionally hummocky and includes enormous blocks of previously-solidified lava and tephra, often tens of metres across, the largest being 120 m. This is a debris avalanche deposit, evidence of a sudden huge collapse of the volcano's north side, following its emergence on 20 January 1731 (ie Phase 1). The sudden reduction in pressure on the rising magma triggered a strong explosive eruption, causing shock waves experienced in Gran Canaria. Large bombs were carried northwards for hundreds of metres and an ash cloud settled over most of the island, from Yaiza to Haría. Curbelo vividly describes these extreme events (wrongly dating them as 10 January): "a mountain of some size was seen to rise up, and on the very same day sink back down into its own crater with dreadful noise, covering the island with ash and stones". This entire sequence took just 7 days.

87. Montaña de las Nueces looking west across its crater towards Rilla and the Señalo/Partido complex

4.2.2. Phase 2: Montaña del Señalo

Phase 2 started in March 1731 when Señalo's eruption added to the cluster of overlapping cones and tephra fields in the central area, adjacent to Partido. At 517 m, Señalo is the highest of the 1730's volcanoes. Several new vents opened until July 1731, making this Partido/Señalo complex Lanzarote's greatest concentration of vents [**88**]. The south-flowing lavas were more viscous than the fluid lavas of Phase 1 because the magma had a different history during its ascent in the upper mantle, so it produced shorter flows of the (blocky) aa type.

88. Señalo/Partido complex

89. Timanfaya complex with (Pleistocene) Diama in the foreground

4.2.3. Phase 3: Hydromagmatic eruptions in the west

In June 1731 eruptions shifted 12 km to the extreme west, marking the start of Phase 3 with submarine hydromagmatic eruptions off the coast near El Golfo. Don Curbelo records that unfamiliar deep sea fish were washed ashore, suggesting that some eruptions were in deep water. The land eruptions produced a string of small cinder cones, with the eruptive centre migrating eastwards along the NE/SW fracture zone. The first cluster was El Quemado, followed by Rajada, a large volcanic complex of international importance because of its wide range of volcanic features. Phase 3 lavas covered a large area of land in the west, reaching the sea [**143**] along a 12 km stretch from Janubio salinas in the south to the edge of Phase 1 lavas to the north. This stretch includes Los Hervideros [**90**], where a thick lava flow cooled slowly to produce columnar jointing, visible in the cliffs of this narrow inlet. Waves are funnelled into a complex system of caves and blow-holes, made accessible for the public and spectacular on a stormy day! As a result of ongoing wave action, in February 2021 an 8-metre wide hole appeared beside the LZ-703 road near Los Hervideros, exposing the cavity beneath the lava crust which had resulted from drained lava, so typical of Lanzarote's lava flows.

90. Los Hervideros

91. Janubio lagoon bar

92. Caldera Colorada

The NE/SW eruptive fissure is clearly revealed by the four strongly-aligned cones of Calderas Quemadas, formed to the east of Rajada [**78**]. These lie within the Timanfaya National Park at the western end of the route followed by the dedicated tourist coaches. The Quemadas lavas flowed not only north-westwards but also southwards towards Yaiza, destroying Don Curbelo's San Juan chapel in Yaiza in January 1732, along with adjacent farmland. These later Phase 3 lavas were viscous, similar to those of Phase 2, so they didn't reach the coast.

Prior to Phase 3, the enclosed lagoon now feeding Janubio saltworks was a large open bay, containing a harbour for vessels including fishing and general inter-island transport. Approaching from the north-east, Phase 3 lavas entered the north side of the bay in 1731, reducing it's size to about 600 m across and isolating it from the open sea. The lagoon depth is now around 3 m, but the level rises and falls with the tides because the entrance bar is permeable [**91**].

4.2.4. Phase 4: Montañas del Fuego and Timanfaya

Early in 1732 the centre of eruptions became stationary around Timanfaya for 4 years. This was Phase 4, a period for which there are no first-hand accounts. The Montañas del Fuego ("Fire Mountains") were formed during this period, the largest and most spectacular volcanic pile of the 1730's episode, shown here with Diama in the foreground [**89**]. These interlocking cones and lava flows surrounded the earlier Pleistocene volcano of Hilario which became an islote on which various visitor facilities have been built. Initial eruptions were Strombolian, producing overlapping cinder cones and creating the large, complex Fuego edifice which rises to 510 m. These were followed by emissions of viscous lava, although at times fluid lava produced some pahoehoe lava. A wide variety of volcanic features was generated here [**101**], as elsewhere in the region, including cinder cones, hydrovolcanic cones, aa and pahoehoe lava flows, lava tubes, lava channels, vesicular lavas (containing gas bubbles), tephra of all sizes, wind-blown ash dunes [**94**] and a range of spectacular surface distortions created when hot lava continues to flow beneath a cooled solid crust [**83**]. Hornitos are easily recognizable, being 2 to 5-metre-high steep-sided cones produced by localised gas explosions. Hornito Manto de la Virgen can be seen (and photographed) from the Timanfaya dedicated tourist coach [**93**].

93. Hornito, Timanfaya

94. Aeolian ash dunes, Timanfaya

4.2.5. Phase 5: Montaña de las Nueces and Montaña Colorada

Volcanic activity ended in March/April 1736 with eruptions at Montaña de las Nueces [**87**] followed by Montaña Colorada, both at the eastern end of the NE/SW fracture. The LZ-56 road passes between them, each cone being about 500 m from the road, opposite each other. The Nueces eruption was rapid, with extremely fluid lava travelling 21 km in 4 weeks and covering 32 km^2, the largest area covered by one volcano in the entire 6-year episode. Some of these lava flows have pahoehoe surface features, but with internal structures akin to the more viscous (but generally faster-moving) aa flows. Each flow advanced rapidly as a single thick unit, unlike most pahoehoe flows (eg Hawaii today) which advance by thin tongues of fluid lava continuously flowing over older ones, cooling to form lobes, followed by lava bursting out and forwards again. Typical pahoehoe surface features of Nueces include broad lobes rising several metres above the general level, caused by fluid lava lifting and cracking the previously solidified crust.

Most of the Nueces lava first flowed southward and then eastwards, surrounding Montaña Negra. Some lava then continued southwards where it became constricted at Tegoyo, near Conil, as it flowed through a gap (bottleneck) in the ridge of Pleistocene volcanoes and then down slope towards Los Llanitos, near Mácher. In mid-March 1736 another small flow developed northwards towards Tinajo, much of that lava being subsequently buried by the Colorada eruption in April.

The (much larger) eastern flow of Nueces contains a network of linked lava tubes which cease at the Mozaga bottleneck where lava (molten and solid)

95. Colorada bomb

96. Mozaga bottleneck

became constricted by a small, steep-sided ravine. Here the lava flow's solid crust fractured, with slabs becoming aligned in the direction of flow and pushed to the outside of a bend [**96**]. The lava tube sections to the west of Mozaga are named at specific localities, although it is not certain that they are all linked: Cueva de los Naturalistas (1.7 km long), also called Las Palomas [**97**], Cueva de San Bartolomé (105 m) and Las Cuevas de Mozaga (80 m). The thin roof has collapsed in many places to form jameos which are also named: Perrito, Grande, Higuera and Florida. Around the hamlet of El Islote vines are grown in many small collapse structures. Eastwards of the Mozaga bottleneck the fluid lava branched north-eastwards towards Famara and southward towards Arrecife. The Famara branch stopped 4 km from the coast.

The Arrecife branch flowed around the Pleistocene Montaña de Maneje and developed classic pahoehoe surface features in the vicinity of the Manrique Foundation building near Tahíche. It reached the coast at Arrecife's Los Marmoles harbour.

The final eruption of the 1730-36 episode produced Caldera Colorada [**92**], a cinder cone located 2 km north of the first cone at Cuervo. Colorada developed classic lava lakes in its crater, feeding northward-flowing lava flows which didn't quite reach the coast. It also generated spectacular volcanic bombs, notably on the eastern side, readily accessible from a roadside car park via a track with interpretation boards [**95**]. The eruptions lasted only 2 weeks, ending on 16 April 1736.

97. Cueva de los Naturalistas

98. Montaña del Clérigo Duarte (Tao)

4.3. 1824 eruption

Islanders had only 88 years of volcanic peace and quiet. Between 31 July and 25 October 1824 three large Strombolian cinder cones, and about 29 smaller vents, emerged along a single 14 km fracture zone orientated NE/SW in the vicinity of Tao and Tinguaton. A dyke was probably intruded along the fracture, the main volcanoes marking the points where most magma emerged. There was no economic damage because this lava flowed over the older 1730's flows. Eruptions were of short duration, so all three volcanoes produced small lava flows, none of them reaching the coast.

The first, highest (550 m) and eastern-most volcano is Montaña del Clerigo Duarte (also called Tao) [**98**]. The second eruption, Nuevo del Fuego (also called Chinero) ran from 29 Sept to 5 October at the western end of the fracture and produced the longest lava flow (7 km). Its fluid lavas show textbook examples of pahoehoe and aa, including a classic lava channel heading to the north-west [**99**].

Tinguaton was the last vent to open, on 17 October. The initial eruption produced several low, overlapping cinder cones, followed rapidly by three lava flows heading in different directions, the longest reaching 5 km. The next morning the main crater was full of water. Later the same day a column of dense black gas arose, as magma and water interacted: a hydromagmatic eruption was in the offing. Increasing gas pressure then caused an explosive eruption which breached the crater rim and a current of hot salty water carved a channel through the day-old lavas. On its way out the jets of water (geysers) left spectacular vertical chasms over 100 m deep, here being explored by a caver [**100**].

99. Chinero lava channel looking south. Chinero cone on the left, behind the building

100. Tinguaton chasm

5. SEDIMENT

5.1. Dating and timing

Sediments have accumulated on Lanzarote ever since it first appeared in Miocene times, around 15 Ma. The older deposits are generally tougher, being better cemented, so they are sedimentary rocks. The youngest sediments remain as loose deposits, the best example being El Jable, the wide, dry, sandy belt between Arrecife and Soo. These contain substances not found in the basaltic rocks, notably fossils and minerals such as quartz and mica which lend themselves to dating methods different from those used with igneous rocks. Many of Lanzarote's sediments are interbedded with basalt lava flows so the two sets of dating techniques can be used in tandem: very useful for checking reliability of results. Healthy disagreement on dates is the norm among researchers and it is rare to find a single agreed age for any given deposit. Fortunately for us the precise date is unimportant, so many sediments are here dated in general terms such as "early Pleistocene" or "mid-Pliocene".

5.2. Lanzarote's oldest visible sediments

At Janubio near-shore marine conglomerates and sandstones containing gastropods and corals have recently been dated at 9.6 Ma (late Miocene). Slightly later (in early Pliocene times) wind blown sand plains and dunes accumulated across the Miocene rocks, often interbedded with lava flows or marine rocks. Representing a former shoreline and dated at 4 Ma, marine and wind-blown sandstones occur around the southern slopes of Ajaches. In the north, pale dune sandstones of a similar age, interbedded with some of Famara's later lava flows, are exposed at three locations near Órzola [**102**]: Valle Chico, Fuente de Gusa and Valle Grande. These have been studied for decades because of their remarkable fossils [**68**]. Much of this sediment was derived from marine deposits which initially accumulated on the sea bed, to be later exposed by lowered sea levels and then blown inland [**113**]. Another source is North Africa from where dust and finer sand was (and still is) blown directly.

101. (Previous spread) West from Montaña Rajada

◀ 102. Dune sandstones near Órzola

103. Insect pods

These Pliocene sedimentary rocks contain the first known land animals on Lanzarote and the first snake found on the Canary Islands (only one vertebra, better than nothing!). They also contain quantities of bird egg shells, tortoise egg shells, numerous land snails and insect (so-called) egg pods. Some of the bird egg shells are thought to belong to the ratite group, large flightless birds like the modern ostrich, but another interpretation is that of a large seabird rather like an albatross. Several species of Shearwater and Puffin were also common.

The insect pods are made of tiny sand grains stuck together to form 2 -5 cm long cylinders with rounded ends and a small hole at one end or occasionally in the middle [**103**]. They have been studied for over a century and variously attributed to bees, beetles and locusts. They occur in vast quantities at several locations across the island, either undisturbed as large nests, or dispersed by weathering and erosion (and vehicles). Their vast numbers in some locations have been interpreted as records of locust invasions ("plagues"): such events occur regularly, sometimes threatening crops. These pods are **trace fossils**: ie not the remains of the creature itself but clear evidence that a creature existed, like footprints and burrows. There has been disagreement over the origin of these small but abundant trace fossils and **a** new scientific name has been proposed, *Rebuffoichnus guanche* (Guanches were the original inhabitants of the Canaries). It is now thought that these cylinders were pupation chambers of a beetle, the rocky covering giving protection to the pupa, the stage between larva and adult.

OXYGEN ISOTOPE RATIOS

Oxygen has 3 stable (ie not radioactive) **isotopes**: ^{16}O, ^{17}O and ^{18}O, the difference being the number of neutrons in the atom's nucleus (that defines an isotope). Almost all the world's oxygen is ^{16}O (about 99.8%), with less than 0.2% ^{18}O. The **ratio of these two isotopes** varies with air and water temperature and is useful for determining past climates. We'll call ^{18}O and ^{16}O "heavy" and "light" oxygen for convenience.

When seawater evaporates the light oxygen turns to water vapour more readily than the heavy, leaving seawater enriched in heavy: so seawater's **oxygen isotope ratio** (OIR) is high (it's conventional to quote heavy first). All marine organisms take in oxygen from this seawater, locking it into their bones, teeth, eggs, shells, soft tissue and so forth. So they keep a record of the ocean's high OIR at the time they died. On the other hand, the evaporated water, with low ^{18}O, moves into the atmosphere as water vapour, some condensing to form clouds and then rainfall, snowfall, rivers and lakes: ie fresh water and ice sheets. Plants and animals which get their oxygen from the air or from fresh water [**104**] end up containing lower OIR than marine organisms. They also keep an OIR record in their bones etc. when they die. So the OIR record in any fossil organism is useful in determining where it lived and what it consumed.

Oxygen Isotope ratios are also useful in unravelling past climates, especially sea and air temperatures. Higher temperatures cause higher rates of seawater evaporation, with more light oxygen evaporated and more heavy oxygen left behind. These enhanced OIRs then get preserved in marine and freshwater organisms, as outlined above, and in ice sheets.

When temperatures are lower than normal, evaporation rates are lower so the OIR difference between marine and freshwater OIR is reduced. The smaller the gap the lower the environmental temperature. Using this important feature of OIR, fossils and ice cores of known ages can be used to describe climate change over the last million years or so.

104. Fossil land snails and insect pods near Soo

Information on climatic and ecological changes has been extracted from the three Órzola sites. One oft-used technique involves the ratio of two oxygen isotopes (Oxygen-18 and Oxygen-16) preserved in fossil bones, shells and egg shells: see ***Oxygen Isotope Ratios*** info box. Oxygen isotopes and carbon isotopes in land snails and bird egg shells provide information about temperatures, soil moisture conditions, diet and so forth. For example, it is known from isotope analyses that the Shearwaters consumed both fresh water and seawater whereas the large (flightless?) birds were entirely land-based. The climate was generally dry with occasional wetter periods when the sand stopped shifting, vegetation became established and soils developed.

105. Possible fragments of raised beach surviving on a Papagayo cliff

106. Salinas de El Río

5.3. Climate and sea level

5.3.1. Climatic and geological changes

The Canaries are vital in unravelling climatic changes over the past 5 Myr for two main reasons. First, they are ideally located in relation to past and present ocean currents and global wind circulation. Second, they possess a superb range of marine and terrestrial (land-based) sediments and sedimentary rocks which contain fossil and mineralogical evidence of past climates, especially true of Lanzarote.

Through Miocene and Pliocene times North and South America gradually merged, as the Panama Isthmus slowly closed and shallowed. Final closure, around 3 Ma, caused a huge change in ocean circulation across the globe, especially the Central and North Atlantic. The northeast-flowing Gulf Stream and North Atlantic Drift became established, transferring heat and atmospheric moisture from the tropics towards northern Europe. The southward-flowing Canaries Current became part of the return flow, a cool current channelled between Lanzarote and Africa. This has a cooling effect on Lanzarote's climate and is responsible for the island's relatively moist air.

Sediments record climates, ocean currents and sea level changes. As noted above, a global ice age started in earnest at the Pliocene/Pleistocene transition around 2.6 Ma, after 30 Myr of global cooling. Global sea levels fell during glacial phases, often by more than 100 m, and rose during the warmer interglacials. Each glacial/interglacial cycle lasted about 40,000 years until 1 Ma, and about 100,000 years since then. Smaller glacial advances and retreats, **stadials** and **interstadials** respectively, also occurred.

107. A young coastline near Jameos del Agua

Past sea levels are tricky to measure, partly because land levels change too! Volcanic islands sink and tilt slowly under their own weight and Lanzarote is no exception. They also rise as erosion removes some of their weight. All of this is affected by the thickness and strength of the underlying crust, which is exceptionally thick under Lanzarote (about 18 km), so its rate of sinking has been less than the global average. We need a few numbers here. The best estimate of Lanzarote's land sinking rate from 1 Ma to 300,000 BP is 1.7 cm per 1,000 years (that's a total land sink of 12 m), but then rising at 0.7 cm per 1,000 years for the past 300,000 years (total land rise of 2 m). Clearly, these figures (12 m downwards and 2 m upwards) are far too small to explain the 70 m sea level rise needed to cut the highest marine terrace and the 60 m sea level fall to explain Túnel de la Atlántida. So we can definitely blame the ice age glacials and interglacials for most of the Pleistocene sea level changes, not Lanzarote's own movements. Some of these sea level changes averaged 2 cm per year and even 6 cm per year during sudden Meltwater Pulses. That's amazingly rapid!

5.3.2. Glacials: low sea levels

Sea levels fell during glacials. During the last glacial maximum, around 20,000 BP, sea levels were 70 - 130 m lower than today's. That explains Lanzarote's famous submarine Corona lava tube, as noted earlier. The north-east coastline between Órzola and Arrieta was several kilometres further towards the sea, the exposed coastal plain playing host to large sand dune systems. Now it is a new, young coastline created after post-glacial sea level rise [**107**]. Lanzarote was a considerably larger island at this time, encompassing La Graciosa and the other Chinijo Islands.

5.3.3. Interglacials: high sea levels

Higher sea levels during interglacials caused inundation of low-lying land, notably large areas of the south, the Rubicón [**108**]. Marine sediments were deposited on wave-cut terraces, including pebbly and sandy beach deposits, shelly limestones and sand dune sediments re-worked by the advancing sea. Typically beige, pale brown or yellow in colour and well stratified (layered), these sedimentary rocks occur across the Rubicón and around the southern coast [**105**]. They are well preserved near the Punta Pechiguera lighthouse [**111**] and inland from Papagayo,

108. Rubicón platform: evidence of high sea levels during interglacials

MARINE ISOTOPE STAGES

More detail is known about the recent geological past compared with the distant, so we need a finer timescale. In 1955 an Italian-American scientist, César Emiliani, developed a scheme using oxygen isotope ratios in marine fossils (foraminifera) of the past 800,000 years. They have been used to build up a fine timescale, subdivided into stages - Marine Isotope Stages, MIS. They are numbered 1 to 21 going backwards from today, with even numbers linked to cold (glacial) phases and odd numbers to warm (interglacial) phases. So, the last glacial maximum is MIS-2 (about 20,000 BP, when Corona erupted) and the double peaks of the last interglacial are MIS-3 and 5, with a mini-glacial (MIS-4) between them. The interglacials of MIS-9, 11 and 13, with their higher sea levels, are relevant to Lanzarote's Piedra Alta limestone.

109. Raised beach, Playa Blanca Marina

110. Six metre marine platform, Pechiguera

for example in the lower sections of a small south-facing valley called Barranco de las Pilas. In places these marine, fossiliferous rocks are interbedded with Pleistocene lava flows. Shallow water and beach deposits, often comprising mixed particle sizes and large rounded pebbles (forming the rock conglomerate) were widespread. Rather deeper water deposits of sand and shell fragments were also laid down on the marine platforms, sometimes overlain by wind-blown sand dunes when dry land returned as sea levels fell. This leaves us today with conglomerates (pebbly rocks), sandstones and limestones containing marine fossils: somewhat surprising for famously volcanic Lanzarote!

Marine terraces are flat, unassuming landforms, the opposite of volcanoes. But they provide crucial evidence of past climates. Also called **marine platforms** or **raised beaches**, they have been formed since the mid-Pliocene warm period and are now preserved as remnants. Most researchers identify about twelve terraces around

LANDSLIDES AND TSUNAMI

Most volcanoes are unstable, liable to collapse at any time for many reasons: increasing weight; slope gradients; earthquakes; subsurface magma activity; dykes; groundwater; and (for coastal volcanoes) sea level change; tidal effects; and wave attack. Oceanic volcanic islands are particularly susceptible to catastrophic slope failure. Entire volcano flanks can suddenly collapse as giant **tsunami-generating landslides**, spreading debris across the ocean floor and leaving a characteristic scar (riscos) on the volcano side matched by a curved embayment in the coastline.

Submarine collapses are also common. Ocean floor surveys around the Canaries reveal many deposits which originated as landslides, avalanches or flows of various kinds. They may represent single catastrophic events, repeated small collapses or continuous flow events over centuries or millennia.

A large catastrophic landslide can generate a tsunami and it might also trigger renewed volcanic outpourings. Several large landslide/tsunami events on Tenerife, La Palma and El Hierro occurred around 540,000 BP, possibly triggered by a rise in global sea levels during an interglacial period (MIS-15). Higher sea temperatures during these warm interglacials may have caused an increase in rainfall and increased lubrication within volcanoes, making them even weaker than usual.

Finally, the spectacular Famara Crags resulted from landslides which started before the Pleistocene Epoch, possibly as a single catastrophic event, but more likely as ongoing erosion over millions of years, punctuated by small or medium landslide events, usually lubricated by water after rainfall. These occur regularly, with a fairly large one on 16 February 2021. Catastrophic or gradual in origin, the Famara Crags remain a stunning sight [**23**].

111. Raised beach, Pechinguera

southern Lanzarote, between 1 and 70 m above present sea level. A descending staircase of terraces has been identified, the oldest (and highest) dated at around 1.2 Ma. Remnants occur on the eastern coastal cliffs and southern slopes of Los Ajaches.

For example, the 30 metre terrace can be picked out across the Rubicón plain, notably on the south east side of Janubio salinas where this level is followed by the road to Playa Blanca. The 6 metre raised beach can be spotted between Pechiguera lighthouse and Janubio salinas where it forms a 5 -15 m wide bench backed by a line of cliffs with caves [**110**]. The 2 m platform was formed during the last interglacial, around 120,000 BP and the lowest one, at around 1 m, by a slightly raised sea level around 7,000 BP during a warm (interstadial) phase early in the Holocene Epoch. Fragments of this lowest platform, with its solid sandstone beach deposit, can be seen near the Playa Blanca marina [**109**] and around the Pechiguera lighthouse.

The fossils found in these marine rocks are abundant, but often fragmented. Researchers can determine seawater temperature by identifying species brought to the island at any given time. Marine snails and bivalves originating to the south spread into Lanzarote seas during warmer conditions: the **Senegalese fauna**. During the cooler periods (but still within interglacials) we find fauna originating from the north, encouraged southwards by lower temperatures and the Canaries Current: these are called the **Mediterranean fauna**.

Analysis of various isotopes (eg carbon, oxygen, uranium) found in fossils can indicate climate change, either directly or indirectly. Corals have been used to date climate and sea level change through the Pleistocene, with specimens collected from various localities including Matagorda, Playa Blanca, Piedra Alta and La Santa.

5.3.4. Piedra Alta: a special case

Piedra Alta is fascinating. On the coast about five kilometres north of Playa Blanca a puzzling conglomeratic (pebbly) limestone layer rests on a palaeosol which in turn lies on a sandy dune layer. This trio is sandwiched between a basalt lava flow dated at 820,000 BP, and a much younger lava flow dated at 200,000 BP. It extends inland for a few kilometres. Clearly our puzzling pebbly limestone must have formed between 820 and 200 thousand years ago.

Since it sits on a terrace 20 m above sea level, it was thought to be a normal interglacial marine deposit, formed during a period of high sea level. Between our two dates we had 7 major interglacials and numerous smaller interstadials, so there was plenty of opportunity. Earlier research narrowed down the date to three interglacials, dated around 500, 420 and 320 thousand years ago, corresponding to Marine Isotope Stages 13, 11 and 9 respectively: see ***Marine Isotope Stages*** info box. Some coral fragments within this enigmatic deposit have been dated around 480,000 BP using uranium isotopes, thereby placing it squarely in the MIS 13 stage, but more recent research on corals suggests 180,000 BP (within MIS 6). Such uncertainty is normal and further research will take us closer to the truth: that is what science is all about.

But this rock has some strange features [**112**]. First, it contains basalt blocks which are largest in the north, getting smaller southwards. They are also angular and jagged, unlike most coastal boulders which are rounded by the waves. Second, the fossils are species that are alive today in both warm and cooler waters, including brachiopods and numerous corals. Third, the fossils come from a range of water depths, from shallow inter-tidal environments to deep water down to 250 m. And many of the fossils are broken fragments, with none in their living positions.

Fourth, the rock grains are poorly sorted (a mixed range of sizes) and they are not well-stratified (layered), unlike most marine platform sediments which are well-sorted and well-stratified by constant wave and current action. All this evidence suggests that this is a **tsunami deposit**, resulting from a nearby catastrophic landslide like the well-documented ones of Tenerife, La Palma and El Hierro: see ***Landslides and tsunami*** info box. If this is the case, then the 20 m altitude may be

112. Piedra Alta

a red herring since a tsunami's upwash can easily reach such heights: this one has been estimated at 45 m. Using the block sizes as evidence, it has been suggested that the tsunami may have been generated by a catastrophic collapse of the Famara Massif to the north. Others have linked it to known landslides in Tenerife, notably the Icod (180,000 BP) and La Orotava (540,000 BP) landslides of northern Tenerife, both occurring during cooler periods with lower sea levels.

5.4. Sand dunes and buried soils

During Pleistocene's glacial phases, wind carried sediment landwards from the exposed sea bed, creating sandy plains and dune systems [**114**], as described above at Famara during the early Pliocene. Intermittently throughout Pliocene and Pleistocene times, these ecosystems developed extensively in the northern half of the island, notably in the El Jable region. Sea bed and Sahara were the two sources for Pleistocene sediment draped across the northern half of the island. Dune systems also developed on the coastal platforms, now lost to the waves.

Over time these sandy deposits have become cemented by calcium carbonate originating from its shell fragments. The oldest are now solid rock, being over 4 Myr old, but many are younger and weakly cemented so are very friable and easily eroded. The dunes became stabilised for extended periods when the supply of sand ceased, allowing vegetation and palaeosols to develop. In places these important **dune/palaeosol sequences** became buried (and often baked) by new lava flows or tephra falls. Drier periods tended to produce dune systems while wetter periods allowed soils to develop, although it's a complex picture.

113. Sand encroaching over Corona lava south east of Órzola

A total of eight dune/palaeosol cycles has been dated between 49,000 BP and 5,000 BP using fossil land snails. These cycles were largely determined by climatic change over a few thousand years, although the initial introduction of shelly/sandy sediment to the island throughout the Pleistocene took place in glacial/interglacial cycles of tens and hundreds of thousands of years. Yes, it's complicated!

To the east of Mala three large dune/palaeosol zones (Jable del Medio) can be seen today [**64**], partly buried by lava from the Guatiza volcanoes around 170,000 BP. Further west, adjacent to Tiagua and on the western edge of El Jable, Montaña Timbaiba [**115**] is a large Pleistocene tephra volcano constructed on top of Pleistocene fossil-bearing dunes which are now exposed. The record in these El Jable sandy deposits shows not only evidence of land-based dune formation during periods of low sea-level but also occasional marine inundation during the interglacial high sea-level conditions. Specimens of the rare marine gastropod Harpa rosea (also called Harpa doris) have been found.

114. Sand dunes near Famara

115. Montaña Timbaiba

5.5. Vegas and Saharan dust

As the world's largest warm desert and largest source of airborne dust, the Sahara has affected Lanzarote's climate and its deposits for nearly 5 Myr. The Saharan sediment, mixed with local airborne volcanic particles, gets trapped within Lanzarote's sediments and palaeosols. Lanzarote's endorheic (enclosed) valleys form effective sediment traps which support productive soils across broad plains (vegas), such as Vega de Guatiza, Vega de Mozaga and Vega de San Jose [**36**] near Teguise.

In winter, Saharan dust is generally carried to Lanzarote by **Calima** winds directly from the Sahara. This low-level airflow (below 1.5 km altitude) results in dust events of various strengths, forming yellow skies and coating vehicles. The occasional big "sandstorms" hit the headlines [**116**]. The Calima event of 22 – 24 February 2020 was the biggest for 30 years, grounding aircraft, creating a serious health hazard and hitting international headlines. Research shows raised death rates during such events.

In summer, atmospheric circulation transports dust in a much more tortuous route from North Africa. The higher-altitude Saharan Air Layer (1.5 to 5.5 km altitude) first carries it from the Sahel (on the southern flank of the Sahara) to latitudes north of the Canaries. Then it sinks to lower altitudes (below 1.5 km) where it is carried by the north east Trade Winds to Lanzarote.

Saharan dust has been extensively analysed across Lanzarote as part of the dune/palaeosol sequences described above. It includes (fine sand) grains of quartz which are usually too large to be transported such vast distances. Other minerals include mica and kaolinite (a clay mineral), neither of which occur in Lanzarote's home-grown basaltic rock. The kaolinite was once thought to have been produced by weathering of plagioclase feldspar from the local basalt, but research shows that weathering rates in this arid climate are far too slow: so it must be Saharan kaolinite after all. Modern rates of dust accumulation have been estimated for La Graciosa at about 20 grams per square metre per year which gives a total of 540 tonnes each year for that small island, equivalent to 16,000 tonnes a year across Lanzarote if it accumulates at the same rate.

Finally, climate change over the last 180,000 years can be glimpsed by dating vega sediments, notably using luminescence methods on Saharan quartz grains. It was once thought that soils developed on land during the wetter and warmer periods whereas sand, silt and dust accumulated during drier and cooler periods. However, it's a complicated picture for two reasons. First, glacials and stadials, with their lower sea levels, were generally cool and wet compared with the warmer and drier periods between them. So, although local sea beds were exposed during glacials, some authors suggest that they became vegetated, thereby binding the sediment and restricting its supply to the land. Second, soils developed on Lanzarote during almost all climatic conditions, wet and dry, cool and warm, leading some authors to suggest that the key factor was not climate: it was the cessation of sand supply permitting prolonged vegetation growth.

5.6. Salt and salinas

Salinas (salt pans), are beautiful, distinctive and iconic features of Lanzarote's coastal landscape and cultural heritage. They have been vital for Lanzarote's inhabitants over the centuries and a few remaining salinas still have economic and cultural importance today. Lanzarote has the oldest and the largest salinas in the Canary Islands: El Río [**106**] and Janubio [**117**] respectively. In some ways salt has been almost as influential as (lack of) fresh water in shaping Lanzarote history, but this is not the place to examine the changing role of salt and salinas in wider cultural contexts. However, since salt is a naturally-occurring sediment formed by geological processes, we make a brief visit to its role in the island's geo-heritage.

Salt, sodium chloride, NaCl, is a mineral called **halite** which forms cubic crystals when unconstrained. It accumulates as a sediment which, with time, becomes a sedimentary rock called **rocksalt**. Clay and other substances in the sediment, including micro-organisms, can give it a pinkish colour. Typically it accumulates in warm, shallow marine environments by the constant evaporation of seawater, often trapped in lagoons.

The basic aim of salinas is to get sufficient quantities of highly-saline seawater (brine) into an enclosed pond so that the final stage of crystallisation leaves enough

116. Calima storm

solid salt for efficient and worthwhile extraction. Seawater is channelled into a sequence of ponds of increasing salinity by various ingenious methods and strange contraptions involving pumps, channels and the clever use of gravity. Each salina has its own distinctive methods to harness the wind and sunshine. Most use ponds in a grid pattern. During the early stages the water solution becomes increasingly saline. In the final stage the water is closed off in its lagoon so that no more brine can enter. The remaining water becomes so saline that salt crystals form, ready for extraction and piling into heaps for collection. Clearly Lanzarote's sunny and windy weather is a great asset in salt-making, combined with the strong historical demand for fish preservation before refrigeration.

Before the Sixteenth Century Lanzarote's inhabitants collected salt from shallow, naturally-occurring coastal ponds and lagoons. Originating as a natural salt pan surrounded by mud and a coastal sand bar, Lanzarote's first recognisable conventional salina, El Río, dates from the 1500's, although the site was previously used for centuries on a small scale. It was mapped in 1590. Located at the foot of the Famara Crags below the Mirador del Río [**129**], this is the oldest salina in the Canaries. Although no longer functioning, the layout remains: several large lagoons and a grid of 30-40 small rectangular ponds. The lagoon base is below sea level so it is constantly fed by percolating seawater, especially during high tides. Today it has become a natural **sabkha** (salt-flat) in which constant evaporation and subsurface replenishment cause salt accumulation in a tidal environment.

Before Timanfaya's Phase 3 lavas reached Janubio, the sheltered lagoon was the site of a prosperous port. Reduction of lagoon size and depth in 1731 triggered the natural accumulation of salt in the shallowest zones, later to be exploited by the establishment of saltworks in 1895. This was expanded in the late 1890's and again in 1910 [**117**]. The engineering system is sophisticated, attracting global interest. Production fluctuated, with interruptions during two world wars, the Spanish Civil War and a huge storm of 1935. The peak annual output of 50,000 tons was reached in the 1950's, but then refrigeration ships caused demand to slump. Today it has extremely strong protection for cultural, scientific and historical reasons, with too many official designations to list here!

Nearly thirty saltworks have existed around the island over the years, most of them built or expanded between 1900 and 1940. Many had previously existed on a small scale for decades. Arrecife's 13 salinas have a colourful history, built within easy reach of the town's port, Puerto Naos, to supply the fishing industry. They are particularly important for their architectural features. Today about half have disappeared beneath developments but some layouts can still be seen around the port and along the coast towards Costa Teguise where the coastal road is called Avenue de Salinero. A small salina, still in production and recently restored, can be see at Los Agujeros, east of Guatiza.

6. LIVING WITH VOLCANOES

6.1. Wine, tephra and La Geria

La Geria is a truly unique landscape; once experienced, never forgotten [**118**]. It is a broad valley orientated NE/SW, occupying the lower land between the string of Pleistocene volcanoes to the south-east and the 1730's lava fields and cones to the north-west [**37**]. It received a covering of wind-borne tephra, especially lapilli, from the 1730-36 eruptions which buried the estate of La Geria, surrounded 2 large Pleistocene volcanoes (Chupadero and Diama) and covered fertile farmland. Initially the 2 m thick tephra blanket destroyed all plant life so the residents departed. Within a few years, however, lapilli were being used as an exceptionally effective mulch for vine growing.

Lapilli are known locally as 'picon'. Most are **vesicular**: they contain a network of small holes (**vesicles**) formed as gas bubbles within magma during eruption. The vesicles retain water condensed from the air during Lanzarote's cool nights, later releasing it to the underlying soil and plant roots. This moisture also contains microbes which obtain nutrients from the chemical weathering of the basaltic lapilli. On release with the moisture, these microbes enhance the nutrient supply to the plants.

The agricultural use of tephra from Pleistocene volcanoes, including Corona, was first recorded in 1592, but this practice expanded steadily after the 1730's eruptions, especially when advocated by Bishop Pedro Manuel Dávila y Cárdenas, Bishop of the Canaries from 1731 to 1738. A plaque on the church at Femés mirador, dated 17 February 1733, commemorates his influence. By the late 1730's picon mulching was firmly established as a standard technique, later spreading across the Canaries and the world.

117. (Previous spread) Janubio salinas

◀ 118. La Geria

LANZAROTE AND WINE

"*Lanzarote and Wine. Landscape and Culture*", written by Rubén Acosta and Mario Ferrer, provides a guide to the wine growing culture of the island. It sets wine-making in its historical, cultural and natural contexts and also gives practical information on the wine activities and the numerous bodegas.

Settlement and farming resumed fully after 1736, since when Lanzarote farmers have used tephra as a mulch for crops, including salads, cereals and roots. For example, within the sheltered crater of Montaña de los Dolores at Mancha Blanca [**121**] tephra mulch provides such an attractive landscape that it warrants its own viewing platform (Guiguan).Vine growing also began at this time, the Malvasia variety being particularly successful using this method. Across La Geria, and beyond, the vines are grown as single plants, without a supporting framework, each in its own pit and often sheltered from the wind by a low curved wall on the windward side [**118**]. The 3 m wide pit is dug down to the soil and manure is added prior to vine planting. Tephra then spills downwards to provide a mulch, typically 50 cm thick: see ***Lanzarote and Wine*** info box. Prior to 1730, the soil had developed on Pleistocene lava surfaces, so it was reasonably fertile. Research has shown that soil moisture beneath tephra mulches is nearly double that in soils with no mulch, the contrast being greatest during the driest months. This is a highly successful model of sustainable farming in an arid region, having minimal inputs of chemicals and energy, so new land is regularly brought into vine cultivation.

Today tephra is excavated for agricultural and ornamental uses across the island from Pleistocene and 18th century deposits, with evidence of past excavation such as here on Tinamala, near Guatiza [**62**]. Some tephra deposits are quarried on a small scale, as on Montaña de Maneje, near Tahíche. Most tephra is cemented to some extent, either by caliche or by fusion of the particles while still hot following eruption.

119. Caliche protecting tephra, Lomo de Camacho

120. Las Grietas

Quarrying of this strengthened tephra requires heavier cutting equipment and the process leaves steep faces. The thick tephra deposits at Lomo de Camacho and Lomo de San Andres, between Tao and Mozaga, originated from Montaña Tamia, a Pleistocene volcano near Tao. Extensive tephra extraction has resulted in a distinctive landscape, shown here with the underlying aeolian sediment exposed [**122**] and protective caliche on top [**119**]. Black tephra is favoured as an agricultural mulch [**36**] and brown is used more for decoration.

Good sections in consolidated tephra can be examined in many places, notably on the southern slopes of Montaña Blanca, near the LZ-35 road. Running water has eroded spectacular, vertical-sided canyons up to 8 metres deep and 1-3 metres wide in these Pleistocene mixed ash/lapilli deposits [**120**]. These are "Las Grietas", an unusual and fascinating landscape feature. Similarly, the much-visited "Stratified City" comprises surreal wind-sculpted remnant pillars of Pleistocene tephra deposits excavated and abandoned in past decades [**123**]. These abandoned quarries are called roferos, this one sometimes being called "Mataburros". It's located beside the LZ-404 road, 2 km east of Teseguite.

121. Guiguan

122. Lomo de Camacho quarries

123. Stratified City

6.2. César Manrique and CACT

Born in Arrecife on 24 April 1919, César Manrique had a profound and long-lasting impact on Lanzarote art, architecture, landscape planning and tourism. Much has been written about this visionary artist so here I only give a brief summary of his geology-related work.

Manrique was strongly influenced by the natural environment, not least the volcanic landscapes. In his own words, "*the Fire Mountains of Timanfaya are an example where forms, colours and textures are almost never ending. This is why my paintings have an extraordinary similarity with the volcanic landscape*". This landscape influence and his enlightened approach to tourism are reflected not only in his numerous sketches and paintings but also in his unique landscape and architecture projects.

After spending 19 years in Madrid and 2 years in New York, Manrique returned to Lanzarote in 1966 to find an expanding tourist industry threatening landscape and environment. Working with his lifelong friend Jose Ramirez (President of the Lanzarote Council) and other far-sighted residents such as Jesus Soto, he mapped out a sustainable future for Lanzarote tourism which would be sympathetic to the landscape. This included strict controls on development (eg no skyscrapers), the creation of half a dozen unique tourist centres and various public

124. Manrique wind toy

125. Timanfaya camels

art installations, notably large “wind toys” on traffic islands [**124**]. After much lobbying in Madrid by Ramirez, Manrique’s vision was eventually adopted by many influential people and put into practice.

Lanzarote’s Centres for Art, Culture and Tourism (CACTs) are exceptional, reflecting the high value attached to the interconnectedness of these three endeavours. The most obviously geological CACT lies at the heart of the 1730’s eruptions: Timanfaya, designated a National Park in 1974. Visitors are able to view the volcanic landscapes from dedicated coaches following specially-built roads [**130**] or by more ecologically-friendly means of transport [**125**].

The high geothermal heat flow in this area, notably around the Pleistocene islote of Hilario, is spectacularly demonstrated by its ability to ignite brushwood [**136**] and to generate geysers from specially-constructed metal pipes sunk into the hot rock [**128**]. The extreme heat is also experienced directly by visitors when they are invited to handle hot gravel and to peer down a well from which rising hot air is used to cook food for the restaurant.

Jameos del Agua [**74**] and Cueva de Los Verdes [**73**] are Manrique-designed CACTs based on the Corona lava tube described above. Jameos del Agua enables visitors to engage with Manrique’s stunning architectural designs in a serene geological setting.

126. Manrique Foundation

It also houses the main geological research station on the island. Cueva de Los Verdes is less developed than Agua but has stunning cave lighting by Jesus Soto [**73**].

The pivotal discovery that tephra-covered farmland increases agricultural yield is a key feature of Lanzarote farming celebrated at the Monumento Al Campesino, the first of Manrique's centres to be built. The 1730's lava and tephra covered much of the island's most fertile land, so this centre pays homage to the resilience of the farmers.

Manrique's first home was built near Tahíche on the 1736 Nueces pahoehoe lava flow which flowed towards Arrecife. Today it is the spectacular Manrique Foundation, housing paintings and other art works [**126**]. The site was chosen by Manrique because of its linked caverns, a form of spheroidal lava tube, often described as "bubbles", but actually created by the expansion of gas-rich molten lava.

The Cactus Garden is a unique and unforgettable CACT, designed to celebrate local cochineal production. It´s a collection of cacti and other succulents of all shapes and sizes, set in an old quarry among novel installations [**127**]. The paths and steps are constructed using contrasting types of lava surface, notably pahoehoe, and the display areas are mulched with tephra. At several places large rounded volcanic bread-crust bombs provide further reminders of the volcanic setting [**63**].

127. Cactus Garden

128. Geyser

129. Mirador del Río

130. Timanfaya coach

6.3. Lanzarote Global Geopark

The United Nations considers Lanzarote to be so significant geologically that UNESCO designated Lanzarote and Chinijo Islands Global Geopark in December 2015. At the heart of this designation lies the management of a geologically-important landscape by, and for, the community. The emphasis is on education, protection and sustainable development. The Island Council (Cabildo) manages the Geopark activities and their website gives full details of the geology, geology-related tourist attractions, links with European and Global Geoparks and a range of community activities.

The geopark area includes much of the sea bed around the island because of its geological significance: see Panorama 9. Many eruptions were submarine and the sea bed also shows evidence of past landslides. The total area of the geopark is 2500 sq km, with the land area being 900 sq km. The most significant sites of geological interest have been identified as Geosites: see ***Geosites*** info box.

Lanzarote's geology and landscapes are so distinctive that, in some ways, they resemble conditions on Mars, other planets and Earth's moon. Consequently, in 2019 the Lanzarote Geopark and Island Council signed a joint agreement with the European Space Agency to start collaborative training of astronauts in geology and astrobiology. This project allows astronauts to practise many activities, including sampling rocks and soil in an environment similar to that on Mars. Lanzarote is also used by NASA, notably for testing various vehicles for use on Mars. In its search for evidence of Martian water and life, on 18 February 2021 NASA successfully landed its Perseverance Martian rover vehicle, naming its landing site Timanfaya.

6.4. Water supplies

Obtaining water on Lanzarote is tricky because of climate and geology. The island receives less than 150 mm rainfall a year, with high evaporation loss because of the sunshine and persistent wind. Lanzarote's geology hasn't helped either. Volcanic landscapes are highly permeable, so little rainfall survives as rivers and lakes on the surface. It's not surprising, therefore, that Lanzarote is a world-leader in desalination technology.

Maretas matter in Lanzarote, but less so since 1964, the year when Europe's first desalination plant was opened in Arrecife. Before that date most of Lanzarote's water came from maretas, large (often round) storage depressions in the ground, fed from flat roofs and other ingenious rainwater collecting devices. They were vital for survival in Lanzarote's arid climate, along with other smaller local storage tanks called aljibes (cisterns) [**133**]. For centuries water was transported in barrels on camel or donkey carts by the aguadores, but the system often failed, so water was brought in by tanker boats (correillos) from Tenerife and Gran Canaria, much wetter islands.

There are some interesting maretas around. The legendary Great Mareta of Teguise was built in the 15th century on the site of a pond. It was 40 m across, 9 m deep and contained 40,000 m^3 of water. Today its name survives as a market square and children's playpark, the Parque de la Mareta [**131**]. It used to look very different. Another large water storage complex was opened in Arrecife in 1913 after eleven years under construction: Maretas del Estado. Comprising 16 separate cisterns, it has since become derelict and partly destoyed, so in 1997 the Maretas Project was established to develop the space for cultural activities. Another mareta at Costa Teguise was the site of a royal residence, La Mareta, built by King Hussein of Jordan who gave it to King Juan Carlos I. It's now owned by the Spanish state.

131. Image of the Teguise mareta present on a postcard from Lanzarote, *ca*. 1895.
Archive of historical photography of the Canary Islands. Cabildo de Gran Canaria-FEDAC.

132. Mala dam, with columnar basalts in early Pleistocene lavas

In the 1940's a canal was dug across the southern slope of Montaña Blanca to divert surface water into two large underground maretas, each over 50 m long. This large scheme was designed to supply water to nearby towns and villages, including Tías and Arrecife, but was only partly successful and abandoned in the 1980's. A similar scheme with six maretas was developed for neighbouring Monte Guatisea. Both (Pleistocene) volcanoes have caliche crusts, allowing rainwater to flow readily over the surface into the canals.

In order to obtain groundwater held in Famara's lava cavities a 1-km long tunnel was dug into the Famara Cliffs in the 1920's, with a rail track to transport the water out. This was Las Galerias project. Between 1946 and 1953 a water pipe was laid to Arrecife to feed a holding tank from which 50 camels would transport water around the island. Numerous tunnels, bridges, cuttings and embankments were needed, many of which remain.The hospital and Parador Hotel had direct pipe feeds.

There's a dam overlooking Mala [**132**] with an interesting history. Originally conceived by farmers in 1958, the idea was developed during the late 1960's as it became clear that Arrecife's desalination plant was inadequate to cope with Lanzarote's growing population. The plan was to use the reservoir as a source for tanker lorries to distribute water around the island. After a feasibility study in 1972

133. Mareta and Aljibe near El Golfo

its planned capacity was increased six-fold to 1.1 million m^3 because of projected tourist numbers. Construction started in 1975 on the site of a small existing pond. But ... (i) the budget doubled, (ii) they didn't build an access road and (iii) the desalination plant was expanded. So the dam failed totally, never containing more than 20% of its planned capacity. It's still there, blocking a narrow ravine overlooking Mala.

Desalination now dominates Lanzarote's water supply, at two sites - Arrecife and south of Janubio. Other possible water sources are not being ignored. In 2017 the discovery of a large groundwater reserve beneath Timanfaya was reported, possibly able to provide 70% of the island's fresh water. However, much uncertainty remains and further investigations are under way. One big problem with Lanzarote groundwater is salt water contamination, because of the island's low altitude. When fresh groundwater is abstracted it is replaced by seawater migrating laterally in the rocks because there is so little downward replenishment on land by rainfall. Another problem is that this groundwater is likely to have accumulated over centuries and millennia, so its extraction would amount to a form of mining, if natural replenishment doesn't keep pace.

6.5. Oil

Currently Lanzarote has no offshore oilfield. But it nearly had one. During the early stages of Atlantic opening, around 200 Ma, an elongated lowland zone developed in the region of continental splitting. Vast outpourings of basalt lava became interbedded with a variety of sediments (clays, sands, salt), this assemblage of rocks becoming the upper layers of the new ocean crust as Africa drifted eastwards at 2 cm/yr. The ever-widening Atlantic steadily accumulated additional marine sedimentary rocks (sandstones, shales, mudstones, limestones) as it travelled. Some of these, especially shales and mudstones, are known to be likely oil and gas source rocks.

Today these oldest Atlantic rocks form the ocean floor between Lanzarote and Africa. Because of their much lower density, the salt deposits have slowly bulged upwards as huge domes (**diapirs**) up to 5,000 m in height, distorting the overlying rock strata and creating conditions ideal for the accumulation and trapping of oil and gas. In short – it's an oilfield, similar in geology to the North Sea region.

Recent research shows that the salt domes have formed mounds of various shapes and sizes on the ocean floor, up to 10 km across and 375 m high, although most are smaller (**pockmarks**). In places the diapirs have pierced the seabed leading to seepage of oil, methane gas, salt solution (brine), mud and various other gases and particles. The oil reserves are estimated to be able to supply 10% of Spain's oil needs. Spain's oil giant Repsol was granted drilling rights 50 km off the Lanzarote coast in 2014, but within 2 years had decided not to exploit the oilfield, to the delight of most islanders but the dismay of the Spanish government. More recently, following an exploration licence granted by Morocco in November 2019, new offshore oil and gas reserves have been discovered. These are now being further investigated.

7. HAZARDS

7.1. Deep plumbing: rising magma

Lanzarote's greatest potential hazards arise from ascending magma, so we need to know about it. Lanzarote's volcanoes were fed by three large magma bodies (intrusive bodies) which arose from the mantle into the lithosphere, mainly the crust, over the past 70 million years, supplying magma for the building of the island [**135**]. Their bases now lie around 10 –15 km depth, well down within the 18 km-thick crust. At more shallow depths, down to about 4 km, their tops are highly complex, reflecting the magma's final branching routes as it rose upwards and sideways to build the island.

The largest intrusive body is centred on San Bartolomé but this does not correspond with an upland volcanic region today [**137**]. This large intrusive body was the source for most of the Pleistocene volcanoes across the centre of the island, and possibly those around Guatiza. The second intrusive body is centred just off the north-east coast: it supplied the Famara and Corona lavas after its magma had flowed upwards and south-westwards in the lithosphere following a sloping trajectory.

The third intrusive body, in the south-west, is the most complex. It is centred under Las Breñas and is thought to be more than 5 km wide at a depth of 12 km, but it is the upper 4 km which is most interesting. First, unsurprisingly, it supplied the magma for the Ajaches shield volcano during the Miocene Epoch. It also fed some volcanoes during the Pleistocene rejuvenation, with magma flowing northwards and north-eastwards. In particular it fed the Pleistocene volcanoes between El Golfo and Tinajo which today form islotes surrounded by the 1730's lavas. Finally, and again not surprisingly, the Las Breñas intrusive body supplied the magma for the 1730's and 1824 eruptions, after it had tunnelled northwards following routes similar to those of the Pleistocene eruptions. Its top, about 1 km across, lies 4 km beneath Timanfaya, at a temperature of about 900°C, its heat being transferred upwards [**136**] by (ground) water vapour and remnant magmatic fluids.

Geothermal energy can be harnessed to produce electricity by thermoelectric generation, a technology using heat directly, rather than using it to drive turbines with steam. New research at Hilario is exploring the feasibility of such a scheme for Lanzarote. It is all a matter of scale and whether geothermal electricity generation can be sustainable and environmentally friendly at Timanfaya.

134. Cavities below lava crust

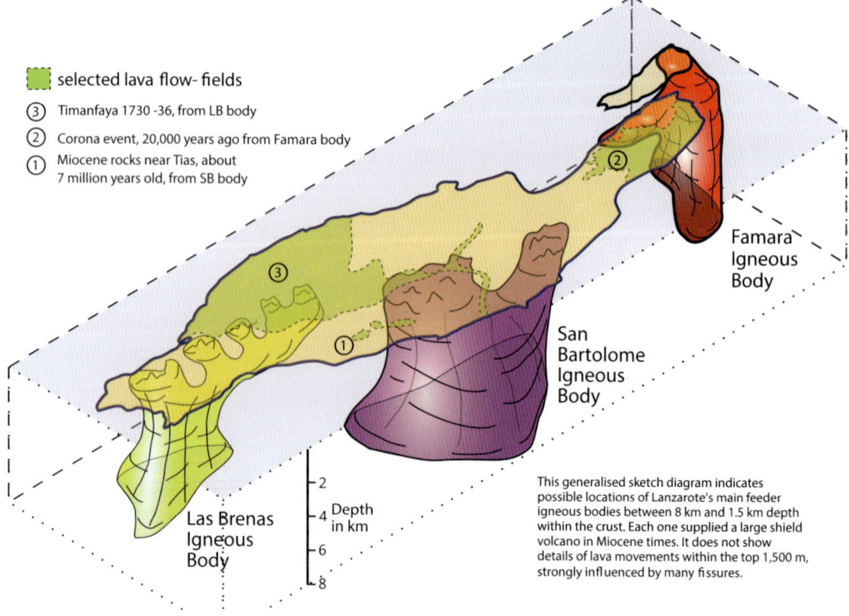

135. Lanzarote: intrusive igneous bodies

136. Timanfaya's remnant heat is sufficient to ignite brushwood

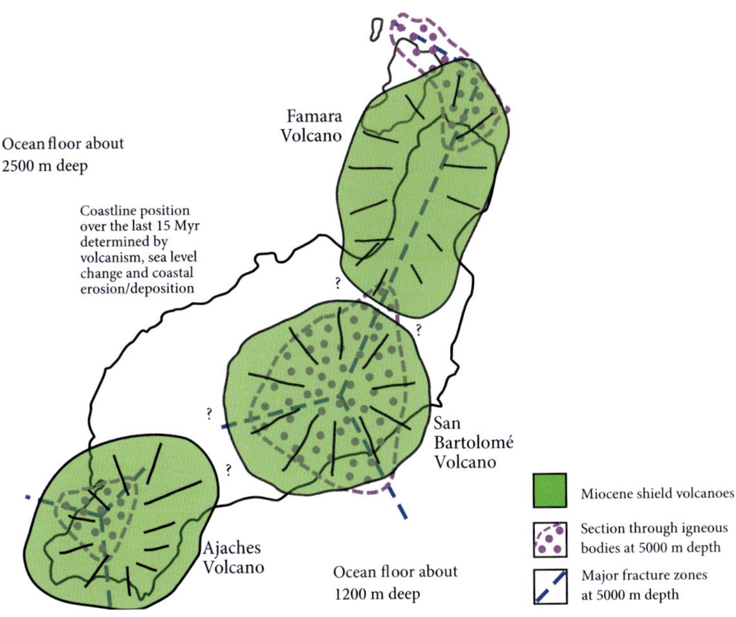

137. Lanzarote: links between Miocene volcanoes and intrusive igneous bodies at 5000 m depth

7.2. Monitoring

Lanzarote has the greatest concentration of volcano-monitoring devices in Europe, and one of the greatest in the world. The island has over 140 sensors across 5 locations, including main surveillance centres at Jameos del Agua, Timanfaya National Park and Cueva de Los Verdes. The heart of this Lanzarote Geodynamic Laboratory is the Casa de los Volcanes at Jameos del Agua, re-opened to the public in 2023 after a major refurbishment. Readings are sent to Madrid every 4 hours, and then onwards to network users. Many monitoring techniques are used. Seismic (earthquake) activity is the most important factor: locations, depths and strengths of the slightest tremors can be picked up by equipment sensitive enough to record earthquakes around the world. Changes in ground surface gradient are also monitored because ascending magma lifts the surrounding rock [**138**]. The island has a network of tiltmeters to detect the slightest change in surface gradient and strainmeters to measure surface deformation (changes in the distance between two fixed points). At Timanfaya tiltmeters and strainmeters are located in two short tunnels 3 m deep which cross each other at right angles so that movement in any direction can be measured.

In addition, GNSS (Global Navigation Satellite System) data are used for constant monitoring of ground deformation. Ground-based equipment monitors gas chemistry, gas temperature, fluid and rock temperature at different depths,

LANZAROTE, A HIKING GUIDE

Written by Ignacio Romero, this compact guide gives a wealth of essential information about Lanzarote's walking routes. It describes 21 self-guided walks ranging from 1.5 to 28 kilometres and includes information about not only route, difficulty, timings and distances but also natural, cultural and historical features along the way.

groundwater levels, gravity strength [**139**], sea level, and magnetic and electrical fields. Changes in geothermal heat flow are especially important because they indicate possible ascending magma. Beneath Timanfaya the magma is still cooling after the 1730's eruptions, with temperatures of around 500° C at 10 m depth and as high as 250° C at a few places on the surface. Ongoing research shows complex relationships between geothermal heat flow, rainfall, water table level, atmospheric pressure and ground deformation. Knowledge of past eruptions is also vital because persistent crustal weaknesses allow new volcanoes to erupt in existing volcanic areas.

Constant monitoring allows the authorities to assess risks. Some relate to the existing volcanic landscapes, especially the subsurface lava tubes and caves, but most risk relates to possible future eruptions, earthquakes and landslides. Lanzarote's volcanoes are dormant, but certainly not extinct: recall that the last Canaries eruption started in September 2021 (La Palma). Regular monitoring and research projects combine to unravel the complexities of Lanzarote's deep plumbing system in order to understand the routes followed by rising magma past, present and future.

138. Surveying ground deformation to detect rising magma

139. Gravity surveying can detect tiny changes in rock type at depth

7.3. Volcanic landscape hazards

Aa lava surfaces are hazardous for walkers because of their sharp edges [7], but greater hazards lie below. Lava tube systems and voids such as volcanic vents and geysers are common in both Pleistocene and historical lava fields across the island, many of them unmapped and representing a potential hazard for walkers and anyone who ventures into the caves [**134**]. The Miocene rocks of Ajaches also contain caves, including a lava tube at Las Breñas found in 2021 to be 2 km long [**140**]and a cave at La Degollada (south of Yaiza) containing archaeological remains. The numerous subsurface cavities across the island attract cavers who survey the tube networks. Others should not venture into the caves without authority: the whole island is a Geopark, most caves lie within protected zones and in many areas it is illegal to leave the footpaths: see ***Hiking Guide*** info box.

Subsurface cavities are also critical for buildings and engineering structures such as bridges because they have the potential to swallow (and waste) huge quantities of poured concrete for foundations. Various techniques have been developed for locating these voids, but no systematic mapping of the island has been undertaken. Ground-penetrating radar is able to locate lava tubes and other voids to an accuracy of a few centimetres, but only to a maximum depth of 20 m.

140. Las Breñas lava tube

7.4. Volcanic eruption hazards

7.4.1. Hazard maps

The main eruption hazards relate to (i) future lava flows, (ii) tephra fallout and (iii) pyroclastic density currents. Monitoring allows the production of hazard maps for use in land-use planning and emergency evacuation schemes. Most of the research aims to identify precursors which might indicate a forthcoming eruption, such as changes in any of the factors mentioned above, especially seismic activity. A second aim is to identify those districts most susceptible to the three types of eruption hazard. It is important work: the 1730-36 eruption had severe economic consequences when the population was only 5,000. Today the island has a resident population of 150,000 and receives over 2 million visitors a year. In essence, hazard maps allow authorities to draw up evacuation plans, although it is clearly much easier to provide information for permanent residents than for visitors.

141. Tajogaite (La Palma) produced lava, bombs and tephra in 2021

Several volcanic hazard susceptibility maps for Lanzarote have been generated recently, the main concern being to locate possible future magma routes to the surface, based on the knowledge that future vents will open near to previous ones. Vent location is also influenced by weaknesses in the rocks, including faults and fissures, many of which persist for millions of years. Dykes might be possible avenues for rising magma, although research shows that pre-existing ones did not act as conduits for magma in 1730 –36.

Generally, Lanzarote basaltic eruptions have a low Volcanic Explosivity Index (VEI), a scale based on the volume of tephra ejected, the altitude reached by the eruption plume and eruption frequency. Hawaiian-type eruptions are usually VEI 1 ("gentle") and Strombolian eruptions are 2 ("explosive"). The largest known eruptions have a VEI of 8, eg Yellowstone 2.1 Ma. Lanzarote eruptions have been taken as VEI 1 or 2, but modern research based on tephra volumes and characteristics indicates a VEI of 4 ("cataclysmic") for some phases of the 1730's eruptions. This new interpretation of the evidence has clear implications for future hazard evaluation and risk management.

7.4.2. Future lava flows

The region at greatest risk from lava flows lies along a NE/SW zone between El Golfo and San Bartolomé: essentially the centre of the historical eruptions [**37**] and one of the Pleistocene chains. Intermediate risk areas are located around Corona and Guatiza volcanoes. An interesting quiet zone (low risk) lies between Teguise and San Bartolomé, believed to result from a Pleistocene cooled magma body at depth which acts as a block to ascending magma, also explaining the lack of historical eruptions in this region.

Maps showing risks from lava flow events are based on (i) likely vent locations (ii) relief, notably lowland routes away from the vents (iii) eruption type, normally assumed to be Strombolian, but possibly more explosive, and (iv) lava characteristics such as viscosity, volume and extrusion rate. Lava flow risk maps based on eruptions on the scale of 1824 are similar to the 1730 –36 maps, but they follow the lower land and are largely restricted to the Timanfaya National Park. These areas lack major infrastructure but include some buildings, roads, utility lines and various tourist facilities. Eruptions larger than the 1824 ones would put at risk towns surrounding the park such as Yaiza, Uga and Mancha Blanca.

7.4.3. Tephra fallout

Tephra fallout from Strombolian eruptions would impact large parts of the island depending on wind direction and the proportions of ash and lapilli, shown here during the 2021 eruption on La Palma [**142**]. Since the prevailing winds are north-easterlies, the most susceptible districts lie to the south-west, including the tourist resorts of Playa Blanca and Puerto del Carmen and the marina of Puerto Calero [**37**]. With other wind directions, the entire island would be at risk, requiring large-scale evacuation.

Volcanic bombs are potential hazards [**141**], but only near the vent. Research on 560 bombs from one of the Timanfaya cones (Caldera Quemada de Arriba) showed that the largest bombs weighed 28 tonnes, being 3 m across with a volume of 10 m^3. These reached 250 m from the vent, the greatest distance (400 m) being reached by 1-tonne bombs.

7.4.4. Pyroclastic density currents

Finally, the greatest hazard is that of pyroclastic density currents resulting from explosive hydromagmatic eruptions. Volcanic material is ejected vertically with force and then falls back as a high-density and high-temperature expanding body of gas, magma droplets and volcanic fragments of all types, flowing downslope at speeds up to 700 km/hour. Little survives. In the past these have been generated not only from pure Surtseyan eruptions (ie offshore), as at El Golfo, but also from intermittent hydromagmatic phases within a normal Strombolian eruption. The risk to the population is extreme and evacuation after such an eruption has started is not usually feasible. That is why constant monitoring of eruption precursors is so vital on Lanzarote.

142. Production of tephra, La Palma, 2021

► (Next spread): 143. Timanfaya's Phase 3 lava flows meet the sea near El Golfo

PLACES TO VISIT: PANORAMAS

Almost anywhere on Lanzarote can provide a stimulating view, so here we suggest a few landscape panoramas, each giving a rich view of a distinctive geo-landscape. They can all be seen from the roadside, or nearby, so no serious hiking is needed. Each panorama is explained briefly in terms of its geology.

1. Haría and Monte Corona from Mirador de Haría, LZ-10, looking north-east.

The large, dark Pleistocene volcano of Corona with its adjacent trio of Quemada de Máguez, Helechos and (much lower) Cerca sit majestically on the Famara plateau, mere 20,000 year-old youngsters compared with the ancient Famara rocks, dissected by gully erosion and weathered to shades of brown over millions of years. Haría and Máguez nestle in the valleys carved in these old Pliocene/Miocene rocks as the original shield volcano was eroded. Mesa de los Llanos is a classic inlier, being entirely surrounded by younger (Pleistocene) rocks. Farmers benefit from the relatively fertile soil which resulted. Parking: Mirador de Haría.

1 Famara plateau: Miocene and early Pliocene basalts and pyroclastics
2 Montaña Quemada de Máguez (La Pescosa)
3 Montaña Helechos
4 Monte Corona
5 Mesa de los Llanos, flat-topped hill in early Pliocene basalts
6 Atalaya de Haría, an early Pleistocene cinder cone
7 Weathered Miocene rocks
8 Old quarries in "Helechos trio" tephra
9 Steep cliffs in Miocene rocks, mainly basalt lavas
10 Haría's southern slopes in Miocene lavas

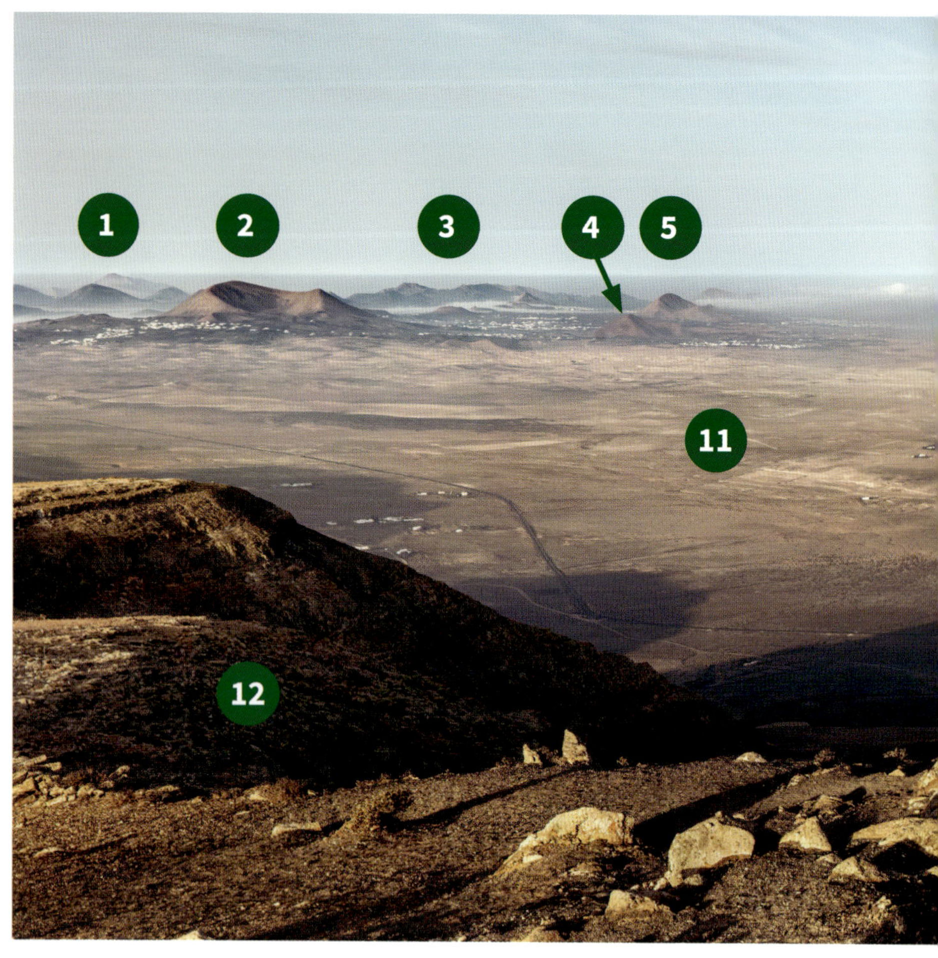

2. Across El Jable towards Soo volcanic chain and Caldera Blanca from Ermita de las Nieves, looking south-west.

This spectacular view shows the Pleistocene Soo volcano chain, orientated NE/SW, surrounded by the wind-blown sands of El Jable, known as the Soo Desert in this northern section. At over 500 m, Montaña Tamia shows the wind-driven asymmetry of many Lanzarote tephra cones. Beneath El Jable's veneer of sandy sediment we find lavas and pyroclastic rocks which originated from the Soo chain and other Pleistocene volcanoes such as those seen in the distance. These rocks are exposed along the narrow coastal zone. In the immediate foreground we see the weathered Miocene basalts which mark the top edge of the dangerously steep Famara Crags. Parking: Ermita de las Nieves

1	Montaña Negra (Pleistocene)
2	Montaña Tamia
3	Timanfaya complex (1730-36)
4	Montaña Timbaiba
5	Montaña Tinache
6	Pico Colorado (Soo)
7	Caldera Trasera
8	Montaña Juan del Hierro
9	Montaña Chica
10	Montaña Cavera
11	Soo Desert
12	Miocene basalts of Famara shield volcano

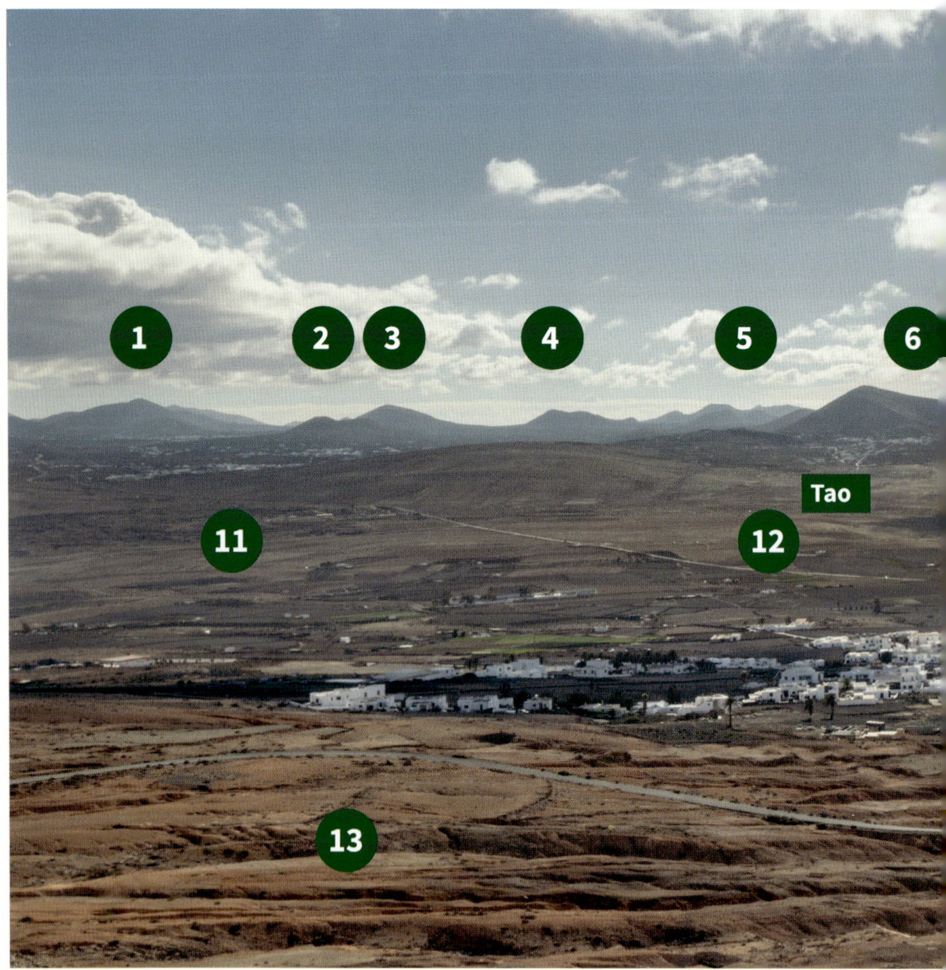

3. El Jable towards Tiagua from Guanapay, looking west.

There's a lot see here! To the far south west we look along the Pleistocene volcanic chain with Guardilama standing proud, with glimpses of ancient Los Ajaches beyond. Then we see some 1730's cones between Cuervo and the Timanfaya massif. Montaña Tamia marks the eastern end of the main Pleistocene volcanic chain through the 1730's flow-fields, ending 20 km to the west at El Golfo. To the north west we have more Pleistocene cones, several appearing to have twin peaks, because of their distinctive asymmetrical shapes. El Jable's sandy plain is interrupted by a very low, wide lobe formed by the release of Nueces lava flowing towards us after it had squeezed through the Mozaga Bottleneck in April 1736. Parking: the top of Guanapay, by the Guanapay Castle.

1	Montaña Guardilama with Ajaches beyond
2	Montaña Diama
3	Montaña Negra (Pleistocene)
4	Montaña Colorada
5	Timanfaya
6	Montaña Tamia
7	Caldera Blanca
8	Montaña Tinache
9	Montaña Timbaiba
10	Montaña Teneza
11	Lobe in Nueces lava
12	El Jable
13	Weathered slopes of Guanapay

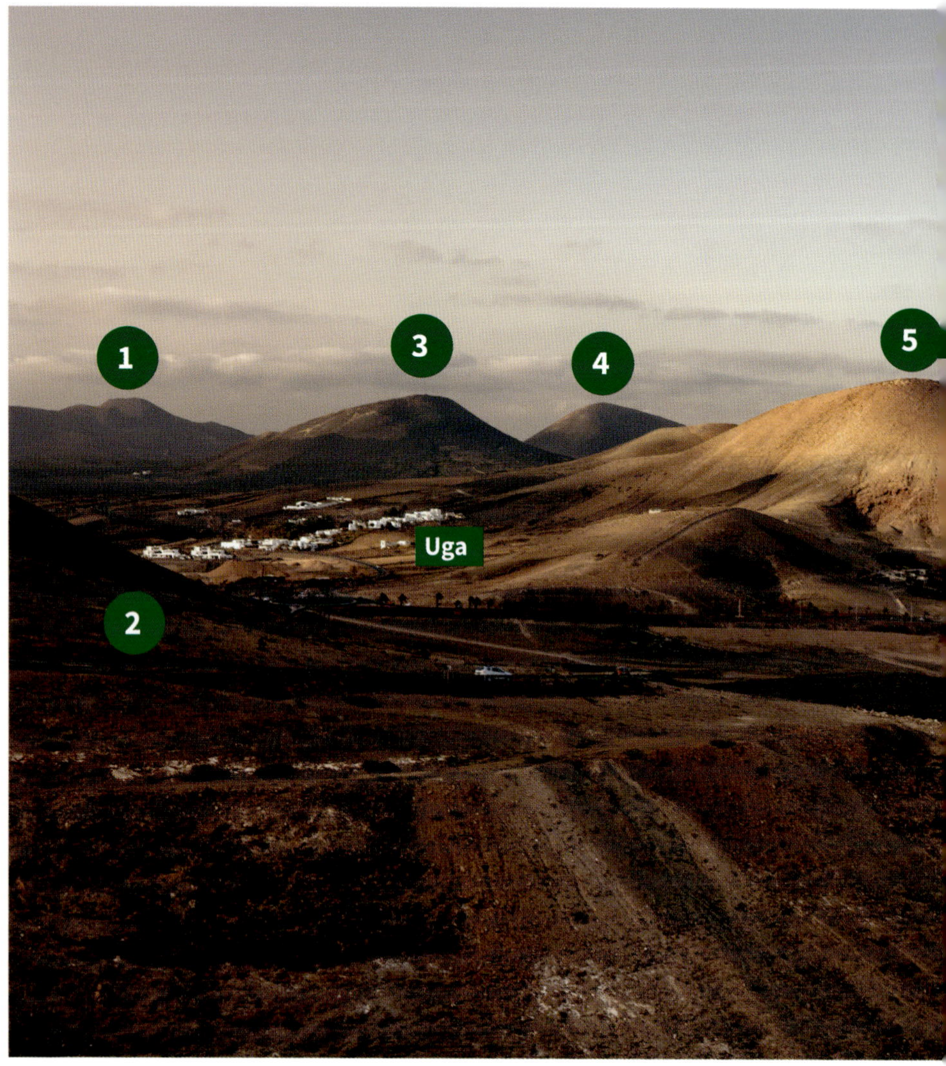

4. Pleistocene volcanic chain from LZ-702 at Las Casitas de Femés, looking north-east.

This classic view shows the major NE/SW line of Pleistocene cones from nearby Riscada to distant Montaña Blanca and beyond. La Geria tephra accumulated in the 1730's between this chain and the two middle-distant Pleistocene cones to the left: Chupadero and Diama. In the far distance, left, we see Montaña del Señalo which erupted in March 1731. The coastal plain running gently down to the coast comprises Pleistocene basalt and pyroclastics, with some small Miocene inliers exposed from beneath the Pleistocene rocks. Riscada's lower slopes occupy the left foreground, dominated by pyroclastic rocks. Parking: Casitas de Femés.

1	Montaña del Señalo
2	Caldera Riscada, lower slopes
3	Montaña Chupadero
4	Montaña Diama
5	Montaña Mojón
6	Montaña Guardilama
7	Montaña Tinasoria
8	Montaña Tersa
9	Montaña Blanca
10	Montaña Bermeja
11	Gentle slopes of Pleistocene lava and pyroclastics

5. Timanfaya and Pleistocene Islotes from LZ-702 at Las Breñas, looking north.

With its patchy soil and vegetation, weathered basalt lava from Atalaya de Femés occupies the foreground, the large mid-Pleistocene composite volcano sitting over our right shoulder. In the centre middle-distance the large islote of two Pleistocene volcanoes, including the double peaked Pico Redondo, is surrounded by black lava from the 1730's events. In the left distance more Pleistocene islotes form the horizon, including the twin peaks of Montaña del Golfo. Beyond Pico Redondo we have two more Pleistocene volcanoes, but further east the 1730's volcanoes dominate, notably the spectacular Montaña Rajada and the massive Timanfaya. The black, fresh, unweathered 1730's lava can be seen flowing towards us from Timanfaya and Rajada. Parking: Las Breñas

1	Montaña Bermeja
2	Montaña Del Golfo
3	Caldera de Chozas with Montaña del Halcones beyond
4	Montaña Quemada (Juan Perdomo)
5	Montaña de Pedro Perico
6	Montaña de la Vieja Gabriela
7	Montaña Hernandez
8	Pico Redondo
9	Montaña Tremesana
10	Montaña Rajada
11	Timanfaya
12	Dark, unweathered lava from 1730's eruptions
13	Weathered beige/brown basalts, mainly from Atalaya de Femés

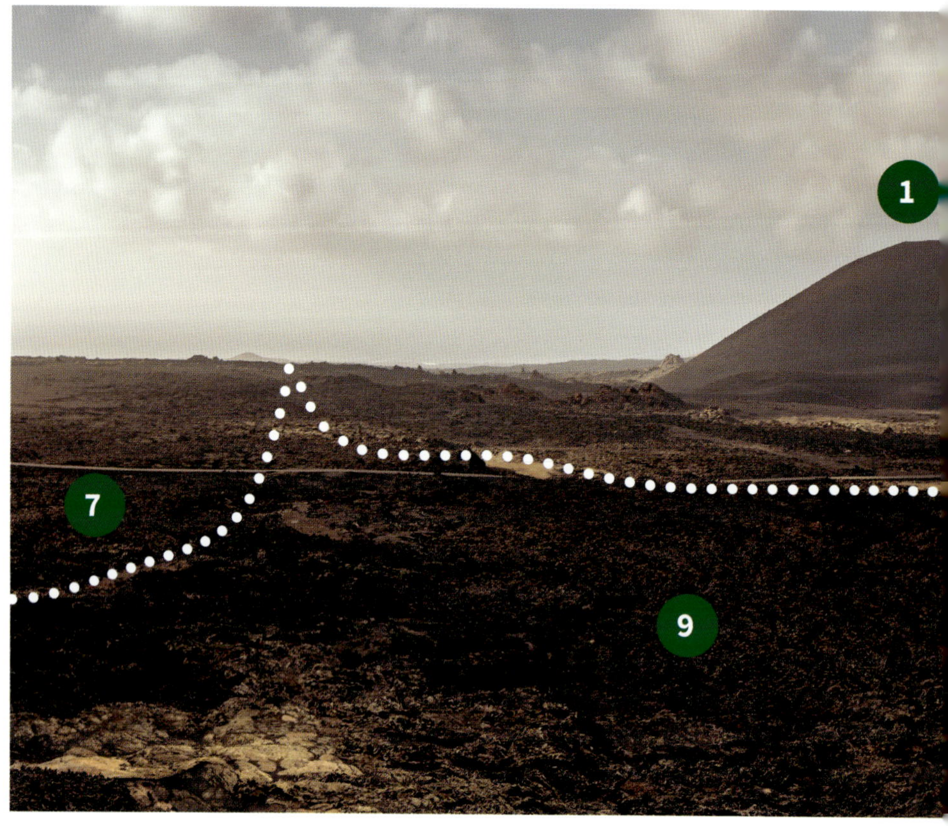

6. Timanfaya lava fields, Mazo & Caldera Blanca from LZ-67/602 junction 1, looking north

Mazo's explosive eruption followed Cuervo's and Partido's earlier lava flows, later joined by Señalo's until June 1731. These extensive Phase 1 lava flows wrapped around the distant and more heavily eroded Pleistocene ash/tuff cone of Caldera Blanca. Partido and Señalo are out of sight to our right. Phase 4 lavas occupy the foreground, having emanated from the extensive and complex Timanfaya/Fuego cluster between early 1732 and early 1736. They sit over our left shoulder. Finally, 29 September 1824 saw the eruption of Chinero, lasting about a week and producing fluid lava that flowed for 7 km directly away from us, to meet the sea at Playa de las Malvas. Chinero is just out of sight to the left.
Parking: about 1 km down the hill at the entrance to Timanfaya CACT.

1	Montaña Mazo (Roja), 20-27 January 1731
2	Caldera Blanca, Pleistocene ash/tuff cone
3	Montaña Teneza
4	Montaña Caldereta
5	Montaña Tingfa
6	Lower slopes of Montaña los Miraderos, a Pleistocene cinder cone
7	Tephra from Chinero
8	Phase 1 lavas, 1730-1731
9	Phase 4 lavas, 1732-1736

7. San Bartolomé from LZ-35 near Montaña Blanca, looking north-east.

Located between Caldera Honda and Montaña Mina, most of San Bartolomé sits on Pleistocene lavas and tephra deposits occupying a gap in the 22 km volcano chain between Uga and Tahíche. Lava from these volcanoes also forms the gentle slopes in the foreground. The LZ-301 runs across a large area of Pleistocene pyroclastic rocks from Monte Guatisea and Montaña Blanca (out of sight to our left). Beyond San Bartolomé the wind-blown sands of El Jable dominate the landscape as far as Guanapay. Montañas Mina and Zonzamas form an upland barrier across this dusty plain. Parking: roadside verge.

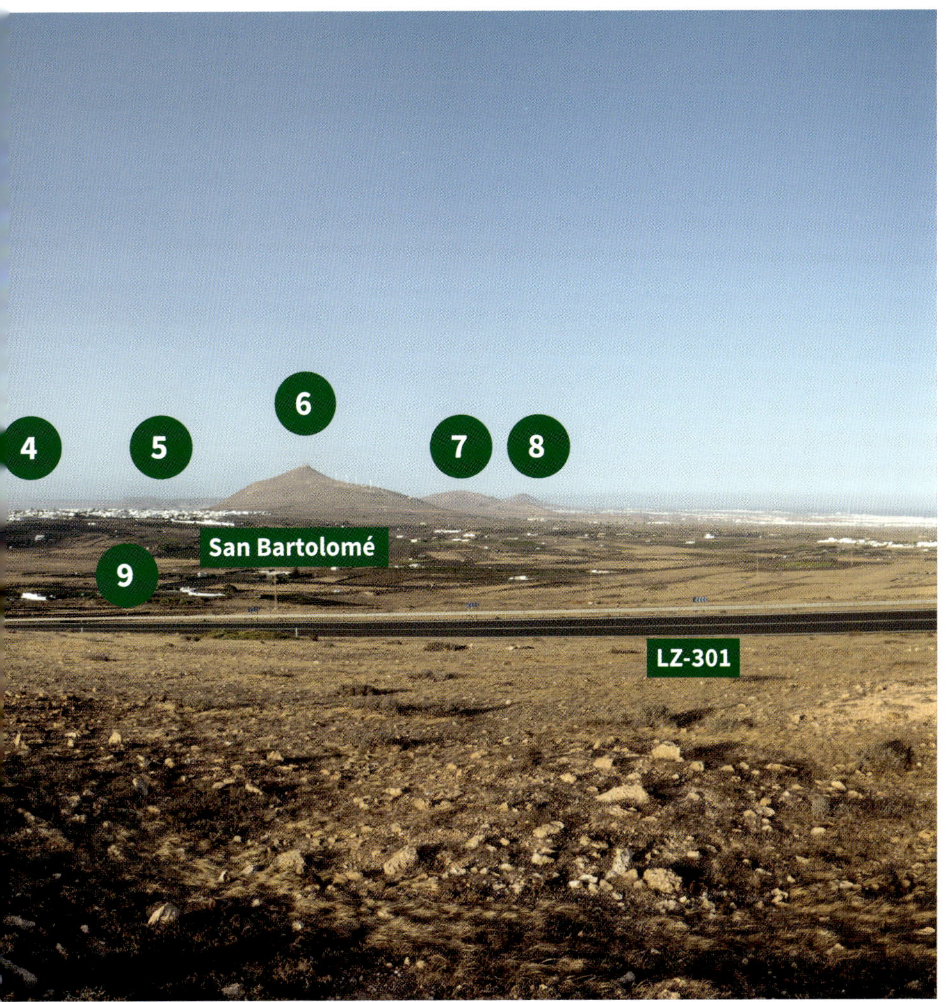

1	Monte Guatisea
2	Caldera Llana
3	Caldera Honda
4	Peñas del Chache, 20 km away, highest point on Lanzarote
5	Guanapay, 13 km away
6	Montaña Mina
7	Montaña Zonzamas
8	Volcan Tahíche behind Montaña de Manaje
9	Pleistocene basalts and pyroclastics

8. Rubicón, Montaña Roja & Ajaches from Femés, looking south-west.

The slope on the left results from erosion over millions of years of Ajaches' horizontal Miocene basalt lavas. By comparison, the lower slopes of Atalaya de Femés (mid-Pleistocene) show beautiful exposures of tephra layers, the gradient determined by the tephra's maximum angle of rest after eruption. Montaña Roja, a large, early Pleistocene composite cone, looms over Playa Blanca. The distant part of the Rubicón Plain comprises Roja lavas, often interbedded with (beige) marine sandstones and limestones, whereas the closer part comprises younger, darker lava and pyroclastics from Atalaya, more irregular and at slightly higher altitude. Parking: Mirador de Femés.

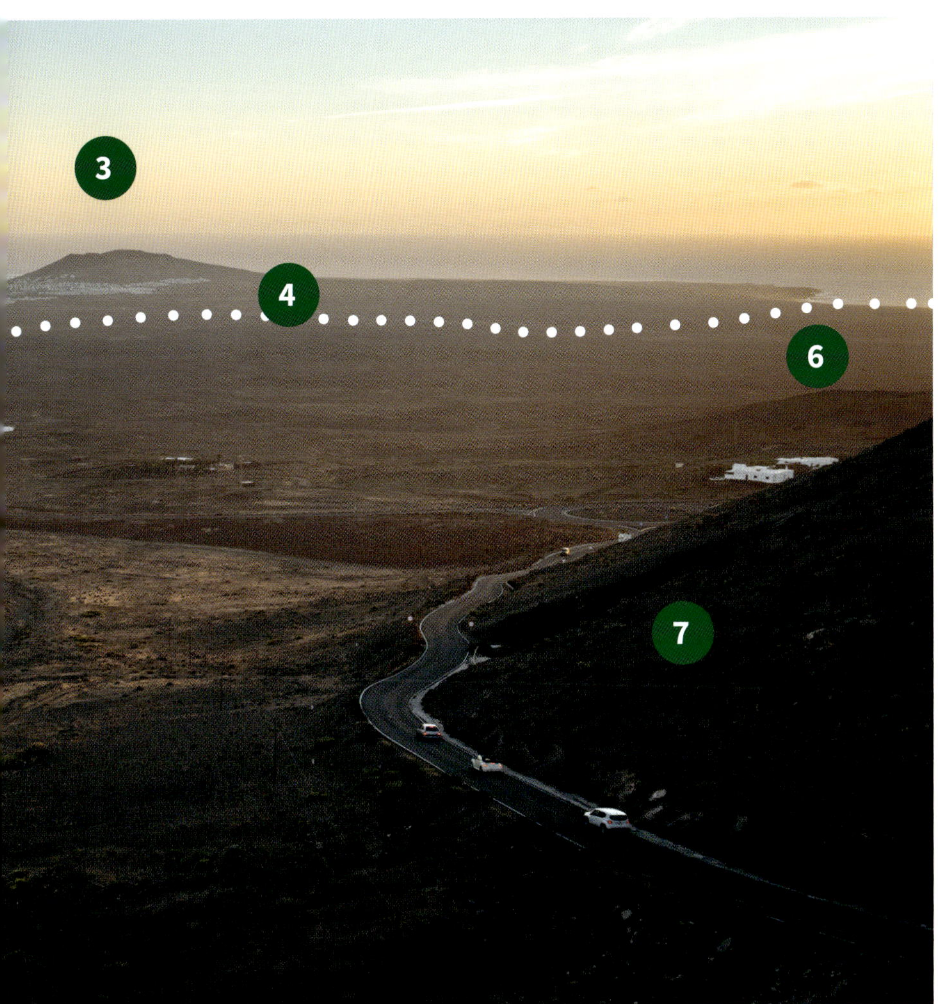

1 Exposed basalt lavas of Ajaches shield volcano
2 Fuerteventura
3 Montaña Roja
4 Distant Rubicón Plain comprising Roja lavas with interbedded sediments
5 Nearby Rubicón Plain comprising dark, irregular Atalaya lavas
6 Caldera de Maciot
7 Lower slopes of Atalaya de Femés with roadside exposures

9. Chinijo Islands from the Mirador de Guinate, looking north

La Graciosa, the largest of the Chinijo Islands, comprises a dozen Pleistocene cinder and tuff cones, aligned NE/SW and surrounded by lava fields and pyroclastic deposits from numerous Surtseyan eruptions. Montaña Amarilla shows an exceptionally wide range of volcanic features and is named (yellow) from its intense palagonitisation (see page 68). Aguja Grande is a typical, wide cinder cone produced by explosive hydromagmatic eruptions. Some of Montaña Clara's submarine feeder dykes are preserved as vertical slabs rising from the sea bed off its south west coast. La Caldera's Surtseyan eruption produced no lava flows at all: the large cone is entirely tephra, with typical features including accretionary lapilli and sag structures. Parking: Mirador de Guinate

1	Montaña Amarilla
2	Wind-blown sand covering Pleistocene lava and tephra
3	Montaña Clara
4	La Caldera (Blanca)
5	Montaña del Mojón
6	Montaña Lobos
7	Montaña Bermeja
8	La Aguja Grande
9	Famara's Miocene basalt lavas

ACKNOWLEDGEMENTS

This book arose from talks I gave to hotel guests in Lanzarote's two Hipotels: Natura Palace and La Geria. Sincere thanks to staff and senior management at both hotels who have been most accommodating and supportive over 2 decades. Lanzarote residents Liz and Larry Yaskiel have also provided ongoing encouragement, allowing me to write articles in Lancelot Magazine, where Liz is Designer and Larry is Editor.

I am heartily grateful to Valentin Troll, Professor of Petrology at Uppsala University and co-author of a major book on the Canary Islands, who generously read through an early draft of the book. He made detailed comments and suggested amendments in a positive and most encouraging fashion. I am also indebted to Maria Jose Huertas, Professor of Geology at the Complutense University of Madrid, for kindly answering my questions about the latest research on Lanzarote geology, for sharing some recent research findings and for taking specific lab photographs. Sincere thanks also to Professors Troll and Carracedo for writing the Foreword.

Members of Lanzarote's caving fraternity have been most helpful in relation to the island's spectacular lava tubes. Carmen Smith and Laurens Smets have been exceptionally generous by providing photographs, regular updates on new discoveries and suggesting improvements to my text.

Text and photographs work closely together in this book: and the photographs are exceptional. Rubén Acosta and Christian Hansen produced most of them and to both I express my sincere thanks, not only for their technical and aesthetic talents, which are obvious in the photographs, but also for their perseverance and patience matching their photographs to my text.

I was able to draw most of the maps and diagrams myself because of the graphics tuition provided so effectively by Ed Oliver, my late son-in-law, who combined immense artistic talent with computer wizardry. Ed kindly constructed pics 4, 8 and 9. I want to thank my grandchildren, Jacob, Emily, Jessica, Camille, Fred and Isaac not only for their patience and understanding when I offer to take them on geology walks but also for the tricky questions they insist on asking! Thanks also to their parents (Lisa & Giles, James and Odette, Emma and Ed) for encouraging them! My wife, Sandra, is far more tactful in matters geological and has been immensely supportive of this project for many years.

Finally, I offer sincere thanks to my publishers: Mario Ferrer and Rubén Acosta who make up Ediciones Remotas. Their vision for this project was clear from the start and it has been a delight to work with them to bring it to fruition. They have shown great patience and understanding throughout, not only over my regular requests for changes, but also by the delays caused by the COVID-19 pandemic. They have also generously accommodated my failure to speak Spanish by undertaking all discussion in English.

Roger Trend

ROGER TREND - BIOGRAPHY

Roger has degrees in geology, geography and science (geology) education from UK universities, his PhD research focusing on geology education. After teaching in UK schools for 20 years, he lectured for a further 20 years in education departments at Sheffield, Exeter and Oxford Universities where he specialised in Earth science education. Roger has given many conference presentations and has edited and written numerous research papers, chapters and books for teachers and fellow researchers. He has also been consultant for about 40 children's books on geology, geography and science. Now retired and living near Exeter, UK, he visits Lanzarote regularly with Sandra and their extended family.

PICTURE CREDITS

The photographs featured in this book are the work of Rubén Acosta and Christian Hansen, most of which were commissioned especially for this publication. Rubén Acosta is a seasoned photographer from Lanzarote who has been honing his craft since 2001 and has exhibited in numerous galleries and photographic events worldwide. He has published monographs of his work, including De-Rendering, La Costa Afortunada and Lanzarote. Tours, Volcanoes and Lentils. Christian Hansen is a designer and landscape photographer from the Canary Islands who provides an authentic perspective and a masterful study of light through his photography.

Rubén Acosta:
Photographs: 1, 3, 6, 10, 11, 12, 15, 16, 17, 21, 24, 25, 26, 28, 31, 33, 36, 40, 41, 41, 44, 45, 51, 55, 58, 61, 62, 64, 65, 66, 70, 71, 73, 74, 79, 88, 89, 91, 92, 93, 94, 103, 104, 105, 106, 107, 108, 110, 111, 113, 114, 116, 117, 118, 119, 120, 122, 123, 124, 126, 127, 128, 129, 130, 132, 133, 134, 136, 141, 142, 143. Panoramic views: 1, 3, 7, 8, 9 and the photograph on pages 180-181.

Christian Hansen:
Photographs: 14, 19, 22, 23, 29, 32, 34, 35, 39, 46, 48, 50, 53, 54, 57, 59, 67, 77, 78, 81, 82, 83, 85, 86, 87, 95, 96, 98, 99, 101, 102, 115, 121, 125. Panoramic views: 2, 4, 5 & 6.

Other original images have been contributed by the following photographers:

Roger Trend
Photographs: 7, 18, 30, 47, 49, 52, 63, 69, 90 & 109.
Laurens Smets
Photographs: 75, 76 & 100.
Jose Arnoso, IGEO, Madrid
Photographs: 138 & 139.
Lisette De Graauw
Photographs: 97 & 140.
Alex Curbelo
Photograph: 84.
Carlos Reyes
Photograph: 56.
Maria Jose Huertas
Photographs: 60.
Mercedes Ferrer, IGME, Spain
Photograph: 112

GOING FURTHER

J.C. Carracedo & V.R. Troll, 2016. ***The Geology of the Canary Islands***. Elsevier. Oxford.
This large book is up-to-date and comprehensive.

Don Andres Curbelo, Priest of Yaiza, (1744). ***When The Volcanoes .... Notes about the occurrences between the years 1730 and 1736***. Editorial Yaiza S.L. Lanzarote.
This small 32-page booklet is Don Curbelo's diary, translated into English and enriched with modern commentary and photographs.

T. Greensmith, 2000. ***Lanzarote, Canary Islands***. Geologists' Association Guide No. 62. GA, London.
This 56-page book is written for geology enthusiasts, but is accessible to all ....

E. Mateo, J. Martinez-Frias, & J. Vegas (eds), 2019. **Lanzarote and Chinijo Islands Geopark: From Earth to Space.** Springer.
This wide-ranging book comprises articles covering all aspects of Lanzarote geology ...

Lancelot Magazine
http://www.lancelot.es/prensa.php

Lanzarote Geopark
http://www.geoparquelanzarote.org/en/geoparque-lanzarote-y-archipielago-chinijo/

Satellite-based maps, published at 2 scales by RAI Ediciones
http://raiediciones.es/esp_cartografia.htm

Lanzarote caves
https://www.lanzarotecaves.com/

Lanzarote place names, including volcanoes, with maps (Spanish)
https://toplanzarote.ulpgc.es

La Palma 2021 eruption
https://volcano.si.edu/volcano.cfm?vn=383010

SOURCES

Information has been drawn from many printed and online sources in the writing of this book, including books and research articles far too numerous to cite. In addition to those identified under "Going Further" and "Acknowledgements", the following summary indicates a selection of sources used as background information for text, maps and diagrams, supplemented by personal fieldwork. The geological processes which have produced the Canary Islands are incredibly complex and way beyond the scope of this introductory book, e.g. in relation to magma generation, volcanic eruption and tectonic movement. Consequently, material presented here is simplified and/or will be superseded as ongoing research generates new knowledge and understanding. Every effort has been made to avoid factual errors and to present an appropriate level of accuracy: all maps and diagrams should be treated as simplified sketches.

Geological Survey of Spain (IGME). Geological maps, website: https://igme.maps.arcgis.com/home/webmap/viewer.html?webmap=44df600f5c6241b59edb596f54388ae4
Google Earth and Google Maps
IDE Canarias, Gobierno de Canarias, website: https://visor.grafcan.es/visorweb/
Lancelot Magazine (English), printed
Lanzarote and Chinijo Islands Global Geopark
Lanzarote Information website: https://lanzaroteinformation.co.uk/
Lanzarote y sus islotes por Agustín Pallarés Padilla, website: https://lanzarote-y-sus-islotes.webnode.es/
Timanfaya National Park (Gobierno de Espana), website: https://www.miteco.gob.es/es/red-parques-nacionales/nuestros-parques/timanfaya/
Union Internationale de Spéléologie (UIS) Commission on Volcanic Caves, newsletter. https://digitalcommons.usf.edu/uis_volcanic_caves/
Map [68] has been based on many sources, including IGME geological maps and the following:
Tomasi et al, 2022, Inception and Evolution of La Corona Lava Tube System (Lanzarote, Canary Islands, Spain). Journal of Geophysical Research: Solid Earth, 127:1-24
Carracedo J.C et al, 2003, La erupcion y el tubo volcanico del volcan Corona (Lanzarote, Islas Canarias). Estudios Geologica,59:277-302.
Diagram [**27**] is based partly on Fig 2 in **Machin and Torrado, 2005**, Sixth International Conference on Geomorphology. The Island and its Territory: Volcanism in Lanzarote.
A useful recent paper is:
Tomasi I. et al , 2023, Geology of Lanzarote's northern region (Canary Island, Spain), Journal of Maps,online at
https://www.tandfonline.com/doi/epdf/10.1080/17445647.2023.2187717

DISCLAIMER

Mention of any region, location or specific geological locality in the text does not imply that it is safe or legal to visit it. Lanzarote's rough terrain and hot, dry climate can make walking hazardous for the unprepared. Lava fields can be particularly treacherous not only because of razor sharp edges but also because of dangerous cavities beneath the lava. Volcanic cones may look benign from a distance but they cause problems for hikers every year, as does the remote mountainous district of Los Ajaches. Nearly half of the island has legal protection of some kind and many places have restricted access, or walkers are confined to footpaths, for example Timanfaya National Park. Two lava tube lengths are accessible for tourists: Verdes and Agua. All other caves and lava tubes have environmental protection and must not be entered without proper permission. The author and publisher accept no responsibility in these matters.

In collaboration with: